TO OUR DEAREST DAUGHTER: JULIA

JULY 5TH 1995.

LOVE. MOTHER AND

THE WORLD'S BEST

COOKING

IN COLOUR

THE WORLD'S BEST

COOKING

IN COLOUR

MARSHALL CAVENDISH

HOW TO USE THE RECIPES

Approximate preparation and cooking times are given for each recipe, with the following symbols where appropriate:

Ease (!) easy to prepare and cook (!!) requires care (!!!) complicated

Cost (£) inexpensive (££) moderately priced (£££) for special occasions

Freezing (❄) freezes particularly well

Plan ahead (◷) can be cooked ahead of time; this symbol in the steps indicates that the dish can be prepared ahead to this point.

Weights and measures Both metric and imperial measurements are given, but are not exact equivalents; work from one set only. Use graded measuring spoons levelled across.

Ingredients • Flour is plain white flour unless otherwise stated • Sugar is granulated unless otherwise stated • Eggs are medium (EEC size 4) unless otherwise stated • Vegetables and fruits are medium-sized and prepared, and onions, garlic and root vegetables are peeled unless otherwise stated • Butter for sweet dishes should be unsalted • Block margarine may be substituted for butter, although the results may be different in flavour, appearance and keeping qualities.

Cooking Dishes should be placed in the centre of the oven unless otherwise stated.

Calorie counts When the number of servings for a recipe is variable (eg 4-6), the calorie/kilojoule count is given for the minimum number of people served. Optional ingredients are not included in calorie counts.

This edition published in 1994
by Marshall Cavendish Books
(a division of Marshall Cavendish Partworks Ltd)
119 Wardour Street
London W1V 3TD

ISBN 1 85435 714 X

Some of this material has previously appeared in the Marshall Cavendish partwork
WHAT'S COOKING?

Printed in Slovakia

CONTENTS

F O R E W O R D

THE WORLD'S BEST COOKING IN COLOUR is a comprehensive introduction to the many and varied flavours of international food. Full of practical advice and information on cooking techniques as well as exciting recipes and serving ideas, it is an invaluable reference book for all home cooks.

The book explores favourite traditional meals and menus from seven regions around the world: Africa and the Middle East; the Americas; China; East Asia (which covers Japan, the Malaysian peninsula and the many islands of Indonesia, as well as Thailand, Cambodia and Vietnam); Europe; India and Pakistan; and Oceania (which takes in Australia and New Zealand, the islands of the Caribbean, the Philippines and Hawaii).

Individual chapters are devoted to a different nation or region, giving recipes for its most characteristic and popular meals and outlining the specific techniques used in their preparation. Details of the customs and eating habits of each region, its local ingredients and the traditional cooking methods used in its cuisine make fascinating background reading.

With over 150 delicious recipes, each accompanied by a colour photograph and step-by-step instructions, THE WORLD'S BEST COOKING IN COLOUR will appeal to cookery enthusiasts and gourmets of all levels. Helped and encouraged by the step-by-step photographs and cook's tips, even the most inexperienced cook will be able to create a world-class meal.

Colourful Contrasts from Africa

Due to cultural, colonial and physical influences, no other continent in the world can boast such a variety of cooking styles as Africa

Casamance chicken (page 11)

THE VAST CONTINENT of Africa is a melting pot of colour, culture and race. With so many different countries, tribes and settlers from around the world, the best way to take a look at African cooking is geographically, as each area is influenced by climate, local food produce and colonial flavours.

West Africa's coastline is on the Atlantic and includes Senegal, Liberia, Ivory Coast, Ghana and Nigeria. Most of the people work in and are dependant on agriculture. The countries share a hot and humid climate with heavy coastal rains from May to October. West African cooking is colourful and is based on local ingredients like fish, chicken, peanuts and tropical fruits.

Coastal hospitality

Hospitality is the keynote of cooking in coastal West Africa. The women often cook more than is needed in order to be able to offer any relatives or friends passing by some food. Senegal has some of the biggest fish catches and in coastal areas fish can be seen drying in racks in the sun before being exported.

Colonial south

South African food reflects the country's early history rather than the varied nature of the climate and the land. The Dutch settlers used spices brought back from the markets of the East. This has produced delicious lightly curried and spiced dishes which are distinctly South African. The area has a wealth of natural food: sugar, vegetables and a lush variety of fruit as well as rich fishing waters. The African tribes contributed their staple grain porridge to the national diet. The British settlers adopted the old Cape Dutch settlers' dishes which were to become the national dishes of most groups of people. Only the Indians and Chinese kept to their own cookery.

Tribal flavours

The boundary lines between Uganda, Kenya and Tanzania are somewhat arbitrary relics of colonialism. When looking at East African cookery, national differences are less

Senegal fish stew

- ***Preparation: 25 minutes***
- ***Cooking: 1¼ hours***

1kg/2¼lb thick white fish such as haddock or halibut (if sole or flounder is used, roll up the fillets before cooking)
25g/1oz parsley, finely chopped
3 onions, finely chopped
2 small red peppers, one seeded and chopped, the other sliced
salt and freshly ground black pepper
75ml/3fl oz peanut or groundnut oil
3tbls tomato purée
125ml/4fl oz fish stock or water
150g/5oz aubergine, sliced
125g/4oz turnips, sliced
125g/4oz carrots, sliced
½ small cabbage, cut into quarters
1 green pepper, seeded and diced
150g/7oz potatoes, cut into chunks
steamed rice, to serve

- ***Serves 4*** 
- ***870cals/3740kjs per serving***

1 Mix together the finely chopped parsley, 1 onion and the red pepper. Add pepper and salt to taste. Cut a pocket in each piece of fish and fill with the mixture. If you are using flat fish fillets, roll up pieces round the stuffing and hold in place with cocktail sticks.

2 Heat the peanut oil in a large, deep pan and when hot put in the pieces ▶

important than tribal ones. East African cooking is now a hybrid of African, European and Indian cuisine.

Although it is well known for its bountiful wildlife, meat plays a surprisingly small role in most tribal diets, as it is considered too valuable to be consumed as food. Fish on the other hand is a staple of many East Africans with so many lakes and the coral shores of the Indian Ocean which provide some of the world's finest sea food. Plantains are eaten throughout East Africa, as is white maize porridge, whilst the potato has been adopted as a staple. Swahili cuisine is a fascinating blend of the Indian and Arab cooking of the coastal settlers, based mostly on coconut milk, fruit and chillies.

African vegetable market

FRESH FISH

One of Kenya's largest tribes, the Luo, are known as the biggest fish eaters in the country. The fisherman in the tribe check the nets at 3 a.m. They eat the fish freshly cooked with tomatoes and onions, oil and water in a special earthenware pot which is balanced on three stones over the fire. Left-over fish is dried in the sun and taken to the local markets for sale.

of fish. Fry them quickly on both sides to seal the surfaces. Remove the fish with a slotted spoon and reserve. Add the rest of the onions to the pan. Mix the tomato purée with the fish stock or water and add to the pan. Bring to the boil, lower heat and simmer for 4 minutes.

3 Add the potatoes, sliced aubergine, turnips, carrots and the cabbage quarters. Add the green pepper and sliced red pepper to the pan. Add water to cover and salt and pepper to taste. Bring to the boil, lower heat and simmer, covered, for 30 minutes or until the vegetables are tender.

4 Carefully add the pieces of fish and the potatoes to the pan and simmer for a further 20-25 minutes. Carefully remove the cocktail sticks from the fish. Serve this dish with a bowl of steamed rice.

Casamance chicken

This spicy chicken casserole originates in the area around the Casamance river in Senegal. It is very suitable for a barbecue

- ***Preparation: 15 minutes, plus marinating time***
- ***Cooking: 1 hour***

4 large chicken pieces
juice of 3 lemons or limes
salt and freshly ground black pepper
pinch of chilli pepper
4tbls peanut oil
3 large onions, finely chopped
1/4tsp dried thyme
2 bay leaves

- ***Serves 4***
- ***385cals/1615kjs per serving***

1 In a large bowl combine the lemon or lime juice, salt and ground black pepper, chilli pepper and 1tbls oil. Add the finely chopped onions, thyme and bay leaves. Toss the chicken pieces in the mixture and leave to marinate for a maximum of 2 hours or overnight.

2 Heat the grill to medium-high. Remove the chicken pieces from the marinade and pat dry with absorbent paper. Arrange on the grill rack skin side down and grill 10cm/4in from the heat for 5 minutes per side.

3 Heat the remaining peanut oil in a large saucepan or flameproof casserole in which the chicken will lie in one layer. Remove the chopped onions from the marinade with a slotted spoon and fry gently in the oil until transparent. Add the rest of the marinade and cook gently for 5 minutes, stirring occasionally.

4 Put the browned chicken and 4tbls of water in the pan, spooning the sauce over the chicken. Cover and simmer for 45 minutes. Serve immediately with rice.

Cook's tips

To cook the chicken on a barbecue, remove the pieces from the marinade and pat dry with absorbent paper. Cook them on the barbecue for about 15 minutes on each side, brushing from time to time with the marinade. Turn frequently.

Savoury banana fritters

- ***Preparation: 10 minutes***
- ***Cooking: 4 minutes per batch***

2 very ripe plantains, or green bananas
1 small onion, finely chopped
1tbls ground ginger
salt and chilli pepper
peanut or groundnut oil for deep frying

- ***Serves 2-4*** 
- ***330cals/1385kjs per serving***

1 Put the chopped onion in a bowl and add the ginger, salt and chilli pepper.

2 Peel the plantains or bananas and chop them finely. Mash them well in a bowl with a fork. Mix the mashed fruit and onion mixture thoroughly.

3 Heat the oil in a deep-fat frier to 190C/375F or until a cube of bread will brown in 50 seconds. Using a dessert spoon scoop up the mixture and slightly flatten it with your hand. Fry the rounds, a few at a time, until they are brown on both sides, about 4 minutes. The fritters should be cooked but still soft in the middle. Drain on absorbent paper and serve immediately.

Jolloff rice

This spicy chicken and rice dish is served at weddings and Christmas in Ghana, Nigeria and Liberia. There are many versions of it: this one is from Sierra Leone

- ***Preparation: 40 minutes, plus marinating time***
- ***Cooking: 1¼ hours***

2kg/4½lb chicken, cut in 6 serving portions
2 garlic cloves
4cm/1½in piece of ginger root, peeled
salt and freshly ground black pepper
1 tomato, blanched, skinned and chopped
1 green pepper, seeded and chopped
325ml/11fl oz vegetable oil
2½ large onions, thinly sliced
15g/½oz flour
100ml/3½fl oz tomato purée
1tsp fresh thyme or ½tsp dried thyme
1 chicken stock cube
450g/1lb long-grain rice
3 potatoes, cut lengthways into quarters
450g/1lb white cabbage
finely chopped parsley, to garnish

- ***Serves 6***
- ***850cals/3570kjs per serving***

1 Skin the chicken pieces. Chop 1 garlic clove and grate a 10mm/½in piece of ginger root then crush them both with freshly ground black pepper in a mortar. Rub this paste well into the chicken pieces with your fingers. Cover the dish lightly with cling film and leave overnight in the refrigerator to absorb the flavours.

2 Put the tomato, chopped pepper, remaining garlic clove, and remaining piece of ginger root, grated, in a blender and purée until smooth. Reserve 2tbls of this mixture to flavour the rice, and the rest for the chicken.

3 Heat 300ml/½pt oil in a heavy saucepan or flameproof casserole big enough to fry the chicken in one layer. Blot the chicken portions well with absorbent paper. Fry them in the oil, turning them until brown on all sides. Remove from the pan with a slotted spoon and reserve the chicken and oil separately.

4 Fry the onions in 2tbls of the reserved oil in the saucepan or casserole over medium heat until they are transparent and stir in the ginger mixture. Add the flour and cook for 2-3 minutes, stirring all the time. Add 2tbls of the tomato purée, fry for another 3 minutes.

5 Add the chicken pieces to the fried onion mixture and stir well. Add 125ml/4fl oz water, bring to the boil, then lower the heat and simmer the chicken until it is tender, about 20-30 minutes. Remove the pan from the heat.

6 Meanwhile, heat 2tbls fresh vegetable oil in a large saucepan. Fry the garlic mixture reserved for the rice and the thyme in a large saucepan for 2-3 minutes.

7 Add the remaining tomato purée and 900ml/1pt 12fl oz water, salt and the chicken stock cube. Bring to the boil. Wash the rice in 3-4 changes of water to get rid of the starch. Trickle the rice into the water, stir well and then boil uncovered over medium heat until all the water has been absorbed and the surface of the rice appears pitted with holes, about 10 minutes.

8 Reduce the heat to very low, stir the rice once and cover the pan with a tightly fitting lid, or foil and then the lid. Leave until fully cooked, that is, when a grain is pressed between your fingers and there is no hard core, about 7-8 minutes. Taste and re-season if necessary.

9 Meanwhile heat 2tbls of the reserved oil in the frying-pan over fairly high heat. Fry the potato quarters until they are golden brown on all sides. Reserve them in a warm place.

10 Fry the finely shredded cabbage until transparent. Season with salt and pepper to taste.

11 To serve, arrange the rice on a large platter with the chicken pieces in the centre. Arrange the potatoes round the chicken and scatter the cabbage over the top. Sprinkle with freshly chopped parsley before serving.

Swahili chicken

- ***Preparation: 15 minutes***
- ***Cooking: 1½ hours***

2kg/4½lb chicken
fresh herbs, to garnish
For the stuffing:
40g/1½oz butter
75g/3oz peeled ripe plantain, chopped
350g/13oz onions, chopped
2 large garlic cloves, finely chopped
350g/13oz tomatoes, blanched, peeled and finely chopped
3tbls lime juice
25g/1oz raisins
½tsp cinnamon
salt
fresh herbs, to garnish

- ***Serves 6*** 
- ***380cals/1595kjs per serving***

1 Heat the oven to 220C/425F/gas 7. To make the stuffing, melt 15g/½oz of the butter in a large frying-pan over medium heat and add the chopped plantain. Cook for 3 minutes, stirring frequently. Transfer the plantain to a plate.

2 Melt the remaining butter in the same pan over low heat, add the remaining ingredients and cook over high heat until the liquids have evaporated, about 10 minutes, stirring constantly. Remove from the heat and add the plantains.

3 Salt the chicken lightly inside and out. Stuff it and close the opening with a skewer. Put the chicken on a roasting rack in a roasting pan, put the pan in the oven, then reduce the heat to 170C/325F/gas 3.

4 Cook until the chicken is done but still tender, about 1 hour 20 minutes. remove the skewer, transfer to a serving plate, garnish with fresh herbs and serve.

PLENTY PLANTAINS

Plantains are known and eaten throughout East Africa but they are especially associated with the Ugandan kitchen. Boiled, the plantain forms the main starchy part of the Ugandan diet. In Kenya plaintain dishes are often more elaborate and are less common. Dishes based on plantains are often referred to as matoke.

Crisp-coated plaited doughnuts

The secret of the crisp syrupy outside of *koeksisters* is that they are taken straight from hot oil and dipped into ice-cold syrup. This seals the syrup outside and leaves the inside dryish in contrast

- ***Preparation: 40 minutes, plus 2 hours standing time***
- ***Cooking: 30 minutes***

225g/8oz flour
good pinch of salt
1tbls baking powder
1tbls butter
225ml/8fl oz milk, soured with 2tbls distilled or white wine vinegar
vegetable oil for deep frying
For the syrup:
450g/1lb caster or granulated sugar
225ml/8fl oz boiling water
½tsp cream of tartar
1½tsp golden syrup
½tsp ground ginger

- ***Makes about 18*** 
- ***185cals/775kjs per serving***

1 To make the syrup, stir the sugar into the boiling water in a saucepan. Continue stirring until the sugar dissolves. Add the cream of tartar, raise the heat and boil fast for 5 minutes. Remove the pan from the heat and stir in the golden syrup and ginger.

2 Divide the syrup equally between 2 bowls and cool thoroughly. Put them in the refrigerator to chill well.

3 Sift the flour, salt and baking powder into a bowl. Rub in the butter. Add the soured milk and mix to a smooth dough. Leave the dough to stand for at least 1 hour.

COFFEE AND TEA

Around the beginning of the century coffee and tea were introduced to East Africa by the European settlers. These two commodities now play a dramatic role in the economy of Kenya. Strangers always notice that the streets of Nairobi are richly perfumed with the aroma of freshly brewed coffee.

4 Roll out the dough 5mm/¼in thick. Cut it into strips about 5 × 10cm/2 × 4in. Cut each strip lengthways into 3 equal-sized strips leaving one end joined. Plait the strips and press them together at the free ends to join them.

5 Take a bowl of syrup out of the refrigerator. Heat the oil well in a deep frying-pan over medium-high heat. Drop a few doughnuts at a time into the oil, and fry until brown and puffed.

6 Lift them out with a slotted spoon, shake them over the pan, and dip them at once in the chilled syrup. Turn them over once, then drain them in a sieve standing on newspaper.

7 Repeat the process until all the doughnuts have been fried, dipped and drained. If the syrup starts to get warm swap the bowl with the reserved chilled one. Store the doughnuts in an airtight tin.

Serving ideas

Serve the doughnuts with cream, if wished, as a dessert or for afternoon tea. Use any left-over syrup for adding to fruit salads or other desserts.

PEASE PUDDING – HOT

Maize long ago became South Africa's staple grain. Maize or corn cobs are called mealies. Mealie meal, the same product as polenta or cornmeal makes the porridge throughout the country. When it is made to a stiff consistency like pease pudding, it is the African's staple dish and often replaces potatoes as an accompaniment.

Traditions of North Africa

Spices and fruit blend to produce the unmistakable flavour of North African cooking

THE ARABS CALL the western half of the North African shoreline Maghrib, "where the sun sets", but we know it as Algeria, Tunisia and Morocco, three countries which share a stunning mixture of mountain and desert as well as cooking traditions developed over many long years.

Subtlety of spices

There is no plain cooking in North Africa – meat and poultry are flavoured with subtle combinations of sweet and savoury, blends of cumin, cinnamon, cloves and ginger. Less subtle is the characteristic harissa sauce, a blend of crushed chillis, caraway seeds, garlic and oil: it is searingly hot and should be eaten with care.

Tajine and cous-cous

Tajine is the typical North African stew which takes its name from the pot in which it is cooked. In Morocco, it is a rich and delicious stew of chicken or lamb (page 18), but in Tunisia, it is more like an omelette: a mixture of meat and vegetables is mixed with beaten egg before being slowly baked in the oven. Couscous (page 16) is the most famous dish of North Africa: specially processed semolina is steamed and served with meat or poultry and vegetables carefully arranged on a large platter.

Half-moon pasties with egg (page 17)

Bakers in North Africa preparing local bread

Watch out for the chilli-hot harissa which will be served with this!

Fruits and flavours

In North Africa, it is customary to blend fresh and dried fruit with meat and poultry – prunes, apricots, quinces and lemons are widely used in stuffings and sauces. Sweets and puddings are often rich and sticky combinations of dates, almonds, walnuts and sesame seeds. Delicate flower-waters such as orange-flower and rose are used to blend the flavours together.

Half-steamed couscous

- *Preparation: 10 minutes*
- *Cooking: 30 minutes*

450g/1lb medium or fine couscous
salt

- *Serves 6*
- *170cals/715kjs per serving*

1 Put the couscous into a large bowl, fill it with cold water and strain immediately in a sieve, shaking to get rid of as much water as possible. Turn the couscous on to a large platter, spreading it out with a fork to break down any lumps.

2 Fill the bottom pan of a large steamer with water and bring it to the boil. Fit the steamer into place, making sure the steam does not escape from the sides and the water does not touch the bottom of the steamer.

3 Put a layer of couscous in the top of the steamer. When the steam begins to rise, add the rest of the couscous, salt to taste and cook, uncovered, for approximately 30 minutes.

4 Remove the steamer from the heat and sprinkle about 350ml/12oz cold water over the couscous and turn it out on to a platter. You can reserve the couscous up to 6 hours or proceed immediately with the second steaming to serve with the traditional lamb stew (see recipe below).

Cook's tips

If more convenient, you can do the initial steaming the day before and refrigerate.

Lamb couscous

- *Preparation: 35 minutes, plus 8 hours soaking*
- *Cooking: 2½ hours*

Half-steamed couscous (see left)
25g/1oz chick-peas
2-3tbls olive or vegetable oil
1kg/2lb boned breast of lamb, cut into large cubes
1 medium-sized onion, cut into 4 pieces
1tsp chilli powder or harissa
1tsp turmeric
450g/1lb tomatoes, blanched, peeled, seeded and chopped, or canned, chopped tomatoes
1 large carrot, scrubbed and cut into 6 segments lengthways
1 large courgette
2-3 small potatoes

- *Serves 6*
- *695cals/2920kjs per serving*

1 Cover the chick-peas with cold water and soak them for 8 hours or overnight if more convenient.

2 Place the bottom saucepan of a large steamer over medium-high heat. Add the oil, the cubed meat and onion and cook for about 30 minutes, stirring from time to time. Reduce the heat to medium-low if the meat starts to stick.

3 Mix the chilli powder or harissa and turmeric with the seeded, chopped tomatoes. Add the mixture, the drained chick-peas, and 225ml/8fl oz fresh water to the meat and onions. Bring the mixture to the boil and simmer for about 1 hour or until the liquid has reduced.

4 Add the carrot pieces, the whole courgette and potatoes to the pan with enough water to just cover. Bring to the boil, then simmer for 20-30 minutes.

5 Meanwhile, put the half-steamed couscous in the top part of the steamer and fit the steamer into place snugly. Cook the couscous, uncovered, for about 20 minutes, fluffing up with a fork.

6 Turn the couscous into a heated deep serving dish. Spoon off all the surface oil from the stew and dribble the oil over the couscous with some of the stew liquid. Serve the stew and couscous in separate dishes or together.

Deep-fried half-moon pasties with egg

Brik à l'oeuf
This is the most famous Tunisian snack

• *Preparation: 30 minutes*

• *Cooking: 12 minutes*

12 sheets of filo pastry, or 6 sheets of Chinese spring roll pastry, cut into 22cm/8½in circles
oil for deep-frying
melted butter
For tuna fish and whole egg filling:
175g/6oz canned tuna fish, crumbled
6 small eggs
50g/2oz Gouda cheese, grated
4-6 capers, drained and chopped
For fish and potato filling:
175g/6oz monk fish or other firm white fish fillets, cooked
2 medium-sized potatoes, cooked
25g/1oz flat-leaved parsley, chopped
pinch of turmeric
½tsp harissa or red chilli sauce
salt and freshly ground black pepper
1 egg
lemon wedges, to serve

• *Serves 6* 

• *490cals/1890kjs per serving*

1 Put 2 circles of filo pastry on top of each other or take 1 circle of spring roll pastry, if using.

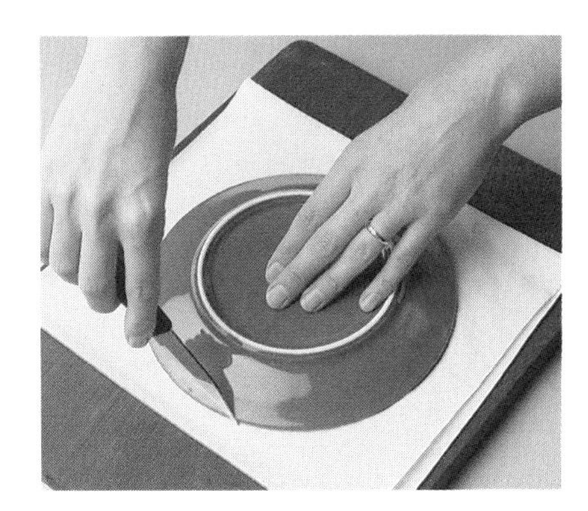

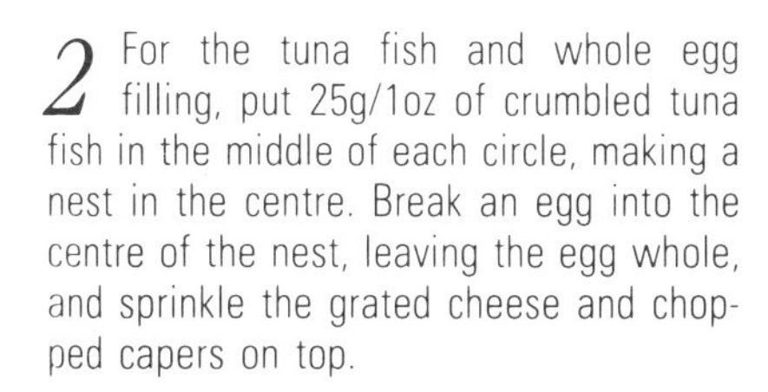

2 For the tuna fish and whole egg filling, put 25g/1oz of crumbled tuna fish in the middle of each circle, making a nest in the centre. Break an egg into the centre of the nest, leaving the egg whole, and sprinkle the grated cheese and chopped capers on top.

3 Brush the edge of the pastry circle with the melted butter. Fold the pastry in half, crimping the sides to seal the edges.

4 Heat the oil in a deep-fat frier to 190C/375F; at this temperature a bread cube browns in 50 seconds. Place the pastry on a fish slice. Carefully lower it into the hot oil and cook until golden on both sides, about 4 minutes, turning once.

5 Drain the pasty immediately on absorbent paper and serve hot. There is an art to eating a brik, as you have to guess where the egg is and bite carefully into the fold. The egg will have a soft-boiled consistency and should be sucked while you nibble the crispy pastry on each side.

6 For the fish and potato filling, mash the fish fillets and cooked potatoes together, lay on the pastry.

7 Add the flat-leaved parsley, turmeric, harissa or red chilli sauce, and salt and freshly ground black pepper to taste. Beat the egg into the mixture.

8 Fold up as shown in the picture above, fry the pasties and serve hot with lemon wedges.

TUNISIAN SPECIALITIES

Tunisia is tinier by far than Morocco and Algeria, but it has a long coastline which produces a great variety of fish. Only in Tunisia will you find the unusual fish cous-cous based on monkfish, sardines and mackerel – and only here will you find that tasty snack, the famous brik, a deep-fried pasty made from thin filo-like pastry and sometimes stuffed with a spectacular whole egg filling (see above).

STEW GENTLY

Most countries bordering the Mediterranean make use of the vegetables they produce, and delicious concoctions are made from tomatoes, peppers and onions, stewed with garlic in olive oil. Add an egg for each person, cook till set and you have Tunisian *chakchouka*. Sprinkle with ground cumin to serve.

Chicken tajine

This dish is said to have been brought to Morocco from Andalucia in Spain.

● ***Preparation: 15 minutes***

● ***Cooking: 1¼ hours***

1.4kg/3lb roasting chicken, jointed
2tbls olive oil
salt
freshly ground black pepper
½tsp ground ginger
½tsp saffron or 1tsp turmeric
2 medium-sized onion, chopped finely
3tbls finely chopped parsley
6 hard-boiled eggs
25g/1oz butter
125g/4oz blanched almonds

● ***Serves 6*** 

● ***470cals/1975kjs per serving***

1 Put the chicken joints into a large saucepan with the oil, salt, pepper, ginger, half the saffron or turmeric, onions and parsley. Cover with water, bring to the boil and simmer gently, half-covered, for 1 hour, or until the chicken has absorbed the taste of the ginger and saffron, and is well cooked. The sauce should be reduced by over one half.

2 Put a little water in a small saucepan, add the remaining saffron or turmeric and heat gently. When the water is warm, shell the hard-boiled eggs and roll them in the saffron water to colour them.

3 Melt the butter in a frying-pan and sauté the almonds for 3-4 minutes until golden. Remove them from the pan with a slotted spoon and reserve.

4 Turn the chicken out onto a deep serving dish and pour the sauce over it. Arrange the eggs on top, placing them around the pieces of chicken, garnish with the sautéed almonds and serve.

MOROCCAN HOSPITALITY

Few restaurants in Morocco can equal the delicious quality of food eaten in the home. The ritual washing of hands precedes an endless banquet of exquisite dishes which ends with glasses of the sweet mint tea which helps the digestion and lures the diners on to syrupy baklava or the Moroccan speciality 'gazelle horns'; these are pastry cones filled with finely ground almonds.

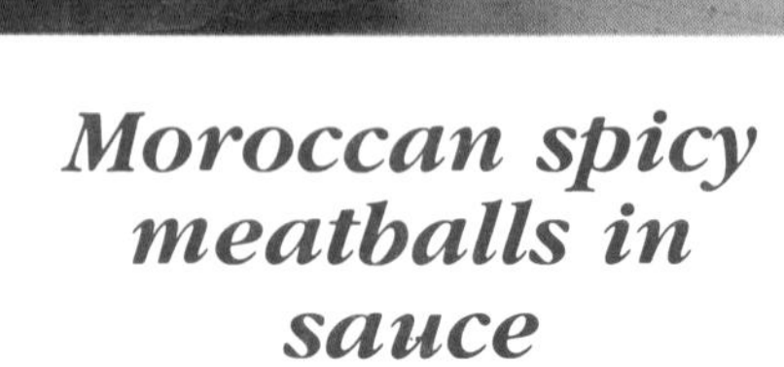

Moroccan spicy meatballs in sauce

● ***Preparation: 20 minutes***

● ***Cooking: 30 minutes***

1kg/2¼lb finely minced lamb
2 onions, grated
2tsp cinnamon
2tsp cumin
salt
3tsp paprika
pinch of cayenne or chilli pepper, to taste
50g/2oz fresh coriander leaves or parsley, chopped
150g/5oz canned tomato purée
juice of ½ lemon

● ***Serves 6*** 

● ***295cals/1240kjs per serving***

1 Place the meat in a large bowl. Add the onions (these can be chopped or grated in a food processor), spices and herbs and mix everything together very thoroughly till you have a paste-like mixture. Roll the mixture into small balls.

2 Meanwhile, fill a large saucepan quarter full of water, add the tomato purée and lemon juice and stir well. Bring the liquid to the boil.

3 Drop the meatballs into the pan, bring to the boil again, then turn down the heat and simmer, covered, until the meatballs are very tender, about 20-30 minutes, and the sauce is reduced. If the sauce reduces too quickly, add a little more water. Turn out onto a serving dish and serve the meatballs immediately.

Spicy fried fish

Here is a dish which is sold in the streets of Tunis and Algiers

- ***Preparation: 20 minutes, plus standing time***
- ***Cooking: 24 minutes***

12 fresh sardines or 6 small red mullet
25g/1oz chopped parsley
2 garlic cloves, finely chopped
½tsp ground cumin
salt and freshly ground black pepper
25g/1oz flour
olive or vegetable oil for frying
1 lemon, cut into wedges

- ***Serves 6*** 
- ***390cals/1640kjs per serving***

1 Gut and clean the fish, leaving the heads and tails on.

2 Mix the parsley, chopped garlic and cumin together. Season with salt and freshly ground black pepper to taste. Using half of this mixture, stuff a little into each fish. Use the remainder to rub into their skins and leave the fish for 30 minutes.

3 Coat the fish in the flour, carefully shaking off any excess.

4 Put ½in oil in a large frying-pan over medium heat. When hot, fry the fish, in batches if necessary, for 2-3 minutes each side, or until cooked. Keep the cooked fish warm. When all are cooked, serve immediately with lemon wedges.

Variations

Larger fish, such as sea bream, grey mullet or sea bass, can be used for this dish. Cut them into 25mm/1in slices and rub each slice with spicy mixture before frying.

OUTSIDE INFLUENCES

The plain and simple food of the original Berber and Bedouin inhabitants has been altered over the centuries by invaders from many lands – the Arabs in the 7th century brought the cooking traditions of the Islamic courts and the Moors returning from Andalucia carried with them the olive oil and grapes of Spain. And of course North Africa fell under the influence of French cuisine when it became an integral part of the French empire in the 19th century.

North African spiced carrots

- ***Preparation: 10 minutes***
- ***Cooking: 15-40 minutes***

1kg/2¼lb carrots, sliced lengthways
1tbls olive oil
2tbls wine vinegar
2-3 garlic cloves
1tsp caraway or cumin seeds, ground
1-2tsp harissa or red chilli sauce
pinch of salt

- ***Serves 6*** 
- ***60cals/250kjs per serving***

1 Put the carrots with the oil, vinegar, garlic, ground caraway or cumin seed, harissa or chilli sauce and salt in a heavy-based pan over a medium-low heat. Cook for 15 minutes, shaking the pan from time to time to prevent sticking. If the liquid dries up, add a little water.

2 Test the carrots after 15 minutes: young carrots will be cooked, but older ones may need up to 40 minutes cooking time. When tender but not soggy, dish the carrots out and serve immediately with a main course or as part of a vegetarian meal.

Cook's tips

The carrots can be mashed and served in purée form as an appetizer.

Algerian chicken with rissoles

Sfereya

Sfereya is an Algerian chicken dish served with bread and cheese rissoles and garnished with whole toasted almonds.

- ***Preparation: 1 hour, plus overnight soaking***
- ***Cooking: 2 hours***

50g/2oz chick-peas
3tbls mixture of butter and vegetable oil
1.4kg/3lb chicken cut into 6 pieces
½tsp ground cinnamon
1 chicken stock cube
freshly ground black pepper
2 large egg yolks
½tsp ground fennel seeds
whole toasted almonds, to garnish ▶

◀ ***For the rissoles***
100g/4oz white breadcrumbs
150ml/5fl oz milk
50g/2oz Gouda cheese, grated
salt and freshly ground black pepper
2tbls olive oil or other vegetable oil

● *Serves 6* 

● ***450cals/1890kjs per serving***

1 Place the chick peas in a large bowl, cover with cold water and soak overnight.

2 Put the breadcrumbs for the rissoles in a bowl, cover with the milk and leave to soak.

3 Meanwhile, put the mixed butter and oil in a large, heavy-based pan over medium-high heat. Add the chicken pieces and sauté them until golden on all sides, about 20 minutes.

4 Add the cinnamon, stock cube, freshly ground black pepper to taste and the drained chick peas. Add 700ml/1¼pt water to cover and simmer until the chicken and chick peas are tender, about 40 minutes. Take the chicken and chick peas out of the pan and place on a warming serving dish. Reserve the liquid.

5 Pour the liquid into a tall jug, wait a few moments for the fat to rise and then skim off the fat. Return the liquid to the pan and boil fiercely until reduced to 225ml/8fl oz.

6 To make the rissoles, squeeze the excess milk from the breadcrumbs. Mix them with the cheese and salt and pepper to taste. Beat in the egg to bind the mixture together.

7 Using 2tsp of the mixture at a time, form it into small, round rissoles about the size of a walnut.

8 Put the oil in a large frying-pan over medium-high heat. When the oil is hot, fry the rissoles until they are golden all over. Drain them on absorbent paper and arrange round the chicken, and keep warm.

9 To make the sauce, beat the egg yolks in a small bowl with the ground fennel seeds. Stir in 1tbls hot chicken stock from the pan. Stir into the remaining stock and cook over low heat, stirring, until the sauce thickens. Do not boil or the sauce will curdle.

10 Pour the sauce over the chicken pieces and rissoles. Sprinkle the dish with the whole toasted almonds and serve immediately.

Orange, date and walnut salad

● ***Preparation: 30 minutes***

1 cos lettuce, washed and shredded
3 medium-sized oranges
50g/2oz dates, (fresh if possible) chopped
50g/2oz fresh walnuts, chopped
For the dressing:
juice of 1 lemon
1-2tbls sugar
2tbls orange-flower water
½tsp cinnamon

● *Serves 6* 

● ***145cals/610kjs per serving***

1 Wash the lettuce and dry with absorbent paper. Place the prepared lettuce in a bowl.

2 Peel the oranges, being careful to remove all the pith, then cut them into medium-thick slices and arrange them in a circle over the lettuce. Add the chopped walnuts and dates.

3 Combine the dressing ingredients in a small jug or jar, pour over the salad, and sprinkle with cinnamon.

Middle Eastern Magic

At the eastern end of the Mediterranean lie Egypt and the ancient lands of the Fertile Crescent, Syria, Jordan and Lebanon, where traditional Arab cooking is still part of the modern way of life.

LOOK CAREFULLY AT the paintings in the tombs of ancient Egypt and you will see the same fruit, vegetables, meat and fish that are sold today in the markets of Cairo and the same simple dishes that are served in the peasant villages on the banks of the Nile. Pictures show cooks preparing the small brown beans called *ful medames* for a rich, long-simmered breakfast stew and chopping up Egypt's most popular vegetable, leafy *melokhia*, the main ingredient for the soup served every day in the Egyptian home.

There is a great difference between peasant fare and the exotic cuisine of the upper class which is an off-shoot of the great traditions of the Islamic Empire where time and money allowed cooks to plan and prepare food of a much richer kind.

Grilled spring chicken with Mixed chopped salad (both page 25)

As in North Africa, spices, herbs and fruit are subtly blended and used to enhance meat stews and stuffed vegetables.

Marvellous mezze

Lebanese cooking is known as 'the pearl of the Arab kitchen' and is

Lunch in the shade under the trees in the heat of the day

certainly the one best known in Europe where many successful Lebanese restaurants have opened in recent years. The most popular item on the menu is almost always the *mezze*, a selection of small dishes so delicious, varied and plentiful that most people fail to travel further down the menu! They are meant to be appetizers, however, tantalising introductions to the traditional main courses of grills and kebabs flavoured with herbs and spices before being skewered for the charcoal grill.

Meatless menus

Good meat is often costly in the Middle East – and pulses are cheap and easy to produce. No wonder, then, that lentils, beans and chick peas are often prepared and cooked in a particularly imaginative and tasty manner. A popular street food in Egypt is the white bean rissole called *ta'amia* – they are flavoured with garlic, onions, coriander leaves and cayenne pepper and are sold hot and spicy from the stall. They are not unlike the falafel of Israel, deep-fried chick-pea balls which are served inside pitta bread.

Beans and lentils are also used to pad out a small amount of meat to feed a large family – Lamb and bean soup (page 24) is a typical example: a mixture of pulses enriches and thickens the broth as the meat slowly simmers.

Stuffed dates

These are as exotic and delicious as they are simple to make. They keep for a long time in an airtight container.

• Preparation: 30 minutes

175g/6oz ground almonds
75g/3oz caster sugar
3-4tbls rose- or orange-flower water
450g/1lb fresh dates

• Serves 8 after dinner 

• 245 cals/1030 kjs per serving

1 Mix the ground almonds and caster sugar in a bowl. Add barely enough rose- or orange-flower water at first to knead them into a firm paste. The oil from the almonds acts as a binding agent when you knead. Add a little more rose- or orange-flower water as necessary.

2 Make a slit on one side of each date with a sharp knife and pull out the stone. Press in a small lump of the almond paste into each date and close it slightly so as to reveal the filling. Store the finished dates in an airtight container. Serve in sweet papers when preparing for special occasions.

Fish balls

- **Preparation: 20 minutes**
- **Cooking: 15 minutes**

1kg/2¼lb cod or haddock or a mixture of other firm white fish
5 x 10cm/4in slices of wholemeal bread, crusts removed
4 medium-sized eggs
2 garlic cloves, crushed
2tsp cumin
salt and freshly ground black pepper
oil for frying

- **Serves 6 as an appetizer** 
- **435 cals/1825 kjs per serving**

1 Cover the fish with water in a saucepan, place it over medium heat and cook always just below simmering for 5 minutes, adjusting the heat as necessary.

2 Drain the fish, remove any skin or bones and flake the meat with a fork.

3 Soak the bread in water, squeeze it dry, then crumble it. Combine the rest of the ingredients, except the oil, in a large bowl, mix well and knead the mixture to a smooth paste. Chill if time permits.

4 Shape the paste into 2.5cm/1in balls. Sauté the balls in hot oil in a frying-pan over medium heat until golden on all sides. Drain on absorbent paper and serve.

Cook's tips

These fish balls can be made in advance and served cold or reheated in the oven.

Yoghurt cheese

- **Preparation: 10 minutes, plus overnight draining**

1.1L/2pt natural yoghurt
1tsp salt
For the garnish
olive oil
1tsp chopped fresh mint leaves
paprika (optional)

- **Serves 6** 
- **100 cals/420 kjs per serving**

1 Beat the yoghurt with the salt, adding more if you like, and pour the yoghurt into a colander lined with a damp muslin or fine cheesecloth. Tie the corners of the cloth together and suspend it over the sink to let the whey drain away overnight.

2 Serve the creamy white cheese in a bowl, garnished with a dribble of olive oil and chopped mint. Sprinkle the yoghurt cheese with paprika.

Cook's tips

You can mix in a variety of raw, finely chopped vegetables such as celery, spring onions, deseeded peppers and chillies before serving.

You can also roll the cheese into little balls and sprinkle them with oil and paprika. Store the yoghurt cheese balls, covered with olive oil, in a jar.

THE MAINSTAY OF LIFE

The annual flooding of the Nile always seemed like a miracle, for it left behind such rich and fertile mud that crops were plenteous and luxuriant. But if for any reason the Nile waters failed to inundate the land, the peasant farmers faced disaster. The building of the Aswan dam has made controlled irrigation possible but the poorer peasants still use the old methods of farming and watering the land.

Stuffed onions

- **Preparation: 35 minutes**
- **Cooking: 45 minutes**

3 large onions
700g/1½lb minced beef
1tbls finely chopped parsley

- *salt and freshly ground black pepper*
- *1½tsp ground cinnamon*
- *½tsp ground allspice*
- *2-3tsp sugar*
- *2-3tsp tamarind paste or the juice of ½ lemon*
- *3tbls vegetable oil*

• Serves 6 as an appetizer 

• 290 cals/1220 kjs per serving

1 Peel the onions and cut off the ends. With a sharp knife make a cut from top to bottom to the centre on one side of each onion.

2 Cover with boiling water in a pan and simmer for 15 minutes or until the onions soften and start to open so layers can be separated. Drain the onion layers in a colander. Cool and carefully separate the layers. The meat may be bought minced but it is better to prepare it at home.

3 When preparing the meat, remove most of the fat. Use a mincer or cut up the meat and chop finely in a food processor.

4 To make the filling, work the meat with the parsley, salt, pepper, cinnamon and allspice. Put a small amount into each onion layer and roll it up tightly.

5 Line the bottom of a heavy-bottomed saucepan with any discarded pieces of onion. Closely pack the stuffed onion rolls seam-side down over the left-over onion pieces in the saucepan.

6 Dissolve the sugar and tamarind paste or sugar and lemon juice in 150ml/5fl oz water and the oil. Pour the mixture over the rolls, adding more water if necessary just to cover them. Place a small plate on top of the rolls, to help the onions keep their shape. Simmer gently over low heat, adding a little water if necessary, for 45 minutes or until the meat is cooked and the water mostly absorbed.

Lamb and bean soup

This nourishing soup makes a substantial winter meal. It can be served with bread and a selection of fresh greens and home-made pickles.

• Preparation: 25 minutes, plus 4 hours soaking

• Cooking: 2 hours

- *50g/2oz dried haricot, navy or other white beans*
- *50g/2oz yellow split peas*
- *50g/2oz black-eyed beans*
- *50g/2oz large brown lentils*
- *500g/18oz stewing lamb, cut into bite-sized pieces*
- *1 large onion, thickly sliced*
- *1tsp cinnamon*
- *½tsp turmeric*
- *salt and freshly ground black pepper*
- *large bunch of parsley, stems discarded and the leaves chopped*
- *4 medium-sized tomatoes, blanched and peeled*
- *1 large aubergine, cubed*

2 medium-sized green peppers, seeds removed and the peppers chopped
4 small potatoes, peeled
bread, to serve
mint, coriander and cress, to serve
spring onions, radishes and pickles, to serve

• Serves 6

• 315 cals/1325 kjs per serving

MILK PRODUCTS

Refrigerators are few and far between in the countryside so it would be difficult to keep dairy produce fresh. What milk there is, is usually turned into yoghurt or butter – and the butter itself will be clarified to preserve it more effectively. In any case, cows and water-buffaloes are used more for farm work than for food production – which is why tender beef is one of the great luxuries in this part of the Middle East.

1 Soak the white beans in plenty of water for 4 hours, then drain them.

2 Put the pulses, meat and onion in a large saucepan with 1.7L/3pt water. Bring the mixture to the boil, lower the heat and simmer for 1 hour, removing the scum as necessary.

3 When the pulses have softened, add the spices and season with salt and pepper. Continue cooking for 1 hour or until the meat is tender.

4 Add the parsley and vegetables to the soup and cook for 30 minutes or until the potatoes are tender. Serve hot with bread and a selection of greens such as mint, coriander and cress along with spring onions, radishes and pickles.

Grilled spring chicken

The best food to buy in Egyptian restaurants is that grilled on charcoal. Giant prawns, chunks of lamb, meatballs, pigeons and young spring chicken are marinated then turned over glowing embers while the delicious aroma entices passers by. Tender chicken is almost as good cooked under the grill.

• Preparation: 5 minutes, plus 1 hour marinading

• Cooking: 20-30 minutes

2 x 1kg/2¼lb chickens, halved or quartered
3fl oz olive oil
juice of 1 lemon
3 garlic cloves, crushed
coarsely ground black pepper
salt
small bunch of fresh coriander leaves or parsley, chopped
lemon wedges

• Serves 4

• 565 cals/2375 kjs per serving

1 Mix the olive oil, lemon juice, garlic and pepper in a bowl or plastic bag. Turn the chicken pieces in this marinade and leave them for about 1 hour or longer.

2 Heat the grill to medium. Drain the chicken, sprinkle the pieces with salt and place them under the grill. Cook for about 10-20 minutes turning the pieces over occasionally and basting them frequently with the marinade, using more oil if they get too brown before they are fully cooked.

3 Place the chicken on a warmed serving dish, sprinkle it with the chopped coriander leaves or parsley, and serve with lemon wedges, pitta bread and a chopped mixed salad.

Cook's tips

Use firm-fleshed white fish steaks such as halibut or monk fish instead of chicken for a tender and flavourful substitute. Turkey breasts are also delicious cooked in this way. A little chilli or hot pepper sauce will add a dash of extra flavour if served separately.

Mixed chopped salad

Add other chopped or thinly sliced raw vegetables such as radishes, white cabbage and carrots to this country salad.

• Preparation: 15 minutes

1 large cos lettuce, finely sliced
3 medium-sized tomatoes, finely diced
1 medium-sized cucumber, finely diced
5-6 spring onions or 1 large mild onion, very finely chopped
1 large bunch of fresh coriander leaves, or parsley finely chopped
For the dressing
6tbls olive oil
2tbls wine vinegar or the juice of 1 lemon
salt and pepper
pepper

• Serves 6

• 145 cals/610 kjs per serving

1 Make sure salad vegetables are properly washed. Drain and dry leaves with absorbent kitchen paper before preparing.

2 Mix the salad ingredients together in a large bowl. Place the dressing ingredients in a screw top jar and shake well. Pour over the salad just before serving.

Cook's tips

If preparing the lettuce in advance, tear the leaves into strips. Slicing with a knife will cause the lettuce edges to brown if it is not served immediately.

Dried fruit salad

Dried fruit was the first foodstuff that the Crusaders brought back from the Orient. In the Middle East dried fruit is used in many savoury, meat and chicken dishes. In Egypt, during the month-long fast of Ramadan a variety of dried fruit is soaked in syrup, sometimes for days, and then eaten when the fast is broken at night.

- ***Preparation: 5 minutes, plus 2 days macerating***

450g/1lb dried apricots (sharp rather than sweet)
225g/8oz prunes
100g/4oz raisins or sultanas
100g/4oz blanched split almonds
50g/2oz pistachios, cut in half, or pine nuts
100-250g/4-9oz sugar
2tbls rose-water or orange blossom water

- ***Serves 6*** 
- ***465 cals/1955 kjs per serving***

1 Put all the ingredients together in a large bowl. Pour in enough water to cover and leave for at least 48 hours, to allow the fruit to plump up and the flavours to blend. Serve the salad with some of its syrup.

Cook's tips

To make a thicker syrup, purée 4-6 soaked apricots with some of the soaking liquid in a blender or through a sieve. Stir this back into the syrup.

HONEY FOR AFTERS

Sweets and pastries heavy with honey are a favourite part of the Middle Eastern menu; they are usually bought in confectioners' shops because they can be fiddly to make at home. Fresh fruit is plentiful – not just the traditional apricots, figs, dates and luscious melons, but newer fruit introduced and cultivated for export as well as home consumption. Oranges, grapefruits, lemons and tangerines all grow in this part of the world now.

A traditional Egyptian dessert, Dried fruit salad (see left)

Southern Comfort

Steeped in history and influenced by the American Indians, Africans, English, French and Spanish, the cooking of the American south is both diverse and unique

Red beans and rice (page 29)

THE SOUTHERN STATES of America comprise a large and varied geographical area, from Virginia and the Carolinas to the deep south as far as Louisiana.

In the early days of the American colonies, the local Indians shared their extensive knowledge of corn production and cooking with the colonists. Many of their corn dishes, such as hominy, have become traditional in the cookery repertoire of the south. Coarsely ground hominy, known as grits, is a common southern breakfast cereal.

As wheat was never successfully cultivated on a large scale in the south, corn is also used to make spoonbread (a kind of baked porridge) and Corn bread (page 31).

The most famous culinary legacies of southern kitchens revolve around seafood, with dishes such as South Carolina's Charleston shrimp pie (page 30) and Maryland's she-crab soup being very popular.

The Africans brought to work on the large tobacco, cotton and rice plantations took with them a love of spices and pilaus, plus okra and sesame seeds, while rich plantation owners used to blend southern cookery with sophisticated English recipes which included trifle and Sally Lunn buns – rich yeast loaves spread generously with butter.

New Orleans specialities

Further south in New Orleans, there is a unique blend of culinary traditions. The French influence is obvious in dishes such as shrimp rémoulade and beignets, while jambalaya was introduced by the Spanish. The native Choctaw Indians contributed filé powder, made of ground sassafras leaves or root, and the black slaves added okra and smoked ham hocks, the hallmark ingredients of Creole-style cooking.

New Orleans is particularly famous for its seafood, especially oysters. Often eaten raw from the half shell, with a little lemon juice or a piquant sauce, they are also made into delicious dishes such as Oysters Rockefeller (page 29). Crayfish are popular, too, boiled in huge kettles and served as they come.

Local women stringing beans in Savanah, Georgia

Mint julep

- ***Preparation: 10 minutes***

1tsp sugar
4-5 fresh mint leaves
crushed ice
50ml/2fl oz or more bourbon
sprig of mint, to decorate

- ***Serves 1*** 
- ***115cals/485kjs per serving***

1 Dissolve the sugar in a few drops of warm water in a chilled glass. Bruise the mint leaves thoroughly and stir them into the sugar mixture.

2 Fill the glass almost to the brim with crushed ice. Add 25ml/1fl oz bourbon and stir to mix the bourbon with the mint-flavoured syrup.

3 Add the remaining bourbon to the top of the glass and decorate.

Variations

Why not try a frosted version of Mint julep? Follow the recipe to the end of step 2, adding the full 50ml/2fl oz bourbon. Dry the outside of the glass and place it in the freezer for 45 minutes-1 hour. To serve, make two holes in the ice with a skewer and insert straws.

Oysters Rockefeller

This elegant dish was invented at the turn of the century at Antoine's, one of New Orleans' most famous restaurants

- ***Preparation: 40 minutes, plus 15 minutes chilling***
- ***Cooking: 10 minutes***

175g/6oz butter, at room temperature
200g/7oz cooked spinach, finely chopped
75g/3oz green part of spring onions, finely chopped
20g/¾oz parsley, finely chopped
25g/1oz celery stalk, finely chopped
½tsp salt
¼tsp pepper
½tsp dried marjoram
large pinch of cayenne pepper
4tbls Pernod
24 large oysters on the half shell
rock salt

- ***Serves 4-6***
- ***440cals/1850kjs per serving***

1 Combine all of the ingredients except the oysters and rock salt, blending well. Refrigerate for 15 minutes or until slightly firm.

2 Heat the oven to 240C/475F/gas 9. Set the oysters on top of a layer of rock salt in large shallow baking tins.

OYSTER TREAT

The abundance of fine oysters has inspired many memorable dishes, but among the simplest is a delicious New Orleans speciality called oyster po' boy (poor boy), made by stuffing a short French loaf with juicy, deep-fried oysters, chopped lettuce and tartare sauce.

3 Place a heaped tablespoonful of spinach topping on each oyster, pressing the topping to the edges of the shell with your fingertips.

4 Bake at the top of the oven for 8-10 minutes or until the topping bubbles and browns slightly. Remove the oysters in their shells from the rock salt and serve immediately.

Cook's tips

You will need about 225g/8oz fresh spinach, before trimming, to give you 200g/7oz cooked spinach.

Red beans and rice

This is a very hearty Creole dish which is loved by rich and poor alike. It is delicious eaten on its own, but it also makes an excellent accompaniment to gumbo

- ***Preparation: 15 minutes, plus overnight soaking***
- ***Cooking: 2¼ hours***

450g/1lb red kidney beans
4tbls bacon fat or oil
1 small green pepper, seeded and finely chopped
1 large onion, coarsely chopped
2 large garlic cloves, finely chopped
2 bay leaves
salt and pepper
225g/8oz smoked ham or pickled pork, diced
boiled rice, to serve

- ***Serves 5-6***
- ***590cals/2480kjs per serving***

1 Soak the beans in water to cover overnight. The next morning, drain the beans and put them in a large pan with 1.5L/2½pt water. Bring to the boil and boil hard for 10 minutes.

2 Meanwhile, heat the fat or oil in a heavy frying pan over medium-low ▶

heat. Add the pepper, onion and garlic and fry them for 6-7 minutes or until the onion is soft but not browned.

3 Add the onion mixture plus the bay leaves to the beans and their cooking liquor and cook over low heat for 1 hour, stirring occasionally. Season the beans to taste with salt and pepper.

4 Add the ham or pickled pork, cover and cook slowly for another hour, stirring occasionally. Place the rice on a warmed serving platter and spoon over the red bean mixture. Serve immediately.

Cook's tips

If pickled pork is difficult to get hold of, make your own by marinating diced raw pork in either distilled or wine vinegar for at least 12 hours in the fridge. Drain and pat dry before using.

A SPANISH FLAVOUR

Jambalaya is the Louisiana adaptation of paella and is a legacy of the Spanish settlement in New Orleans. It can either be made with chicken and meat or with shellfish. Although the basic seasonings are similar to those in gumbo, jambalaya takes on a different character as the rice is cooked with the stew.

Charleston shrimp pie

- *Preparation: 30 minutes*
- *Cooking: 30 minutes*

8 slices of white bread, crusts removed, cubed
350ml/12fl oz milk
450g/1lb cooked, peeled prawns
50g/2oz butter, melted
6 eggs, well beaten
2tsp Worcestershire sauce
4 celery stalks, finely chopped
1 green pepper, finely chopped
large pinch of grated nutmeg
salt and pepper
butter, for greasing

- *Serves 4-5*
- *535cals/2245kjs per serving*

1 Heat the oven to 190C/375F/gas 5. In a large bowl, soak the bread in the milk, then mash it with a fork.

2 Add the prawns, melted butter, beaten eggs, Worcestershire sauce, celery, green pepper, nutmeg and salt and pepper to taste. Stir well to blend all the ingredients.

3 Pour the mixture into a 1.7L/3pt buttered casserole and bake for 55-60 minutes or until the top is golden brown and set. The mixture will still be fairly liquid in the centre. Serve hot.

MORNING CALL

Beignets are a Louisiana tradition. These yeast-raised fritters are often flavoured with lemon zest, brandy or rum. Steaming hot beignets sprinkled with caster sugar are still served with Creole coffee mixed with ground chicory root at a famous New Orleans coffee house called Morning Call.

Corn bread

Corn bread is rather heavy and close-textured, but it makes a delicious tea bread, served in slices and spread with lashings of butter. It is also very popular cut in squares and served instead of rice with other dishes in the American style, such as chilli con carne

- ***Preparation: 15 minutes***
- ***Cooking: 30 minutes***

4tbls oil or melted butter
175g/6oz flour
1tbls baking powder
½tsp salt
175g/6oz fine cornmeal or maize meal
2tbls caster sugar
1 large egg, lightly beaten
300ml/½pt milk

- ***Serves 4-6***
- ***545cals/2290kjs per serving***

1. Heat the oven to 180C/350F/gas 4. Brush an 18cm/7in square baking tin with a little of the oil and line the base with greaseproof paper.
2. Sift the flour, then resift it into a large bowl, together with the baking powder, salt and cornmeal. Stir in the sugar. Make a well in the centre.
3. Pour in the egg, remaining oil and the milk, and mix together lightly but thoroughly with a wooden spoon until a soft dropping consistency.
4. Pour into the prepared tin, smooth the top and bake for 30-35 minutes, or until golden on top and slightly risen.
5. Turn out onto a wire rack and serve warm, cut into squares.

SNOW WHITE RICE

Rice is the foundation of Creole cookery and local cookery books are explicit about the secret of producing perfect rice: snowy white, completely dry, with every grain separate. To achieve this, the rice must be cooked in plenty of boiling salted water for 10-15 minutes or until the grains begin to swell. At this point, the rice must be drained and put in a warm oven for 10 minutes until cooked but still firm. It is then ready to eat.

MUMBO GUMBO

One of the best-known Creole dishes is gumbo. The word gumbo comes from the African word for okra, a common ingredient used to thicken this soup-like stew of which there are many versions, one of the most popular being shrimp. The Cajuns prefer their gumbo with chicken and thickened with filé powder.

Chicken and ham gumbo

- ***Preparation: 25 minutes, plus 30 minutes degorging***
- ***Cooking: 1 hour***

600g/1¼lb aubergines
1tbls salt
4tbls oil
1.2kg/2¾lb chicken, jointed
1 large onion, finely chopped
2tsp paprika
½tsp cayenne pepper
450g/1lb okra
350g/12oz canned sweetcorn, drained
600ml/1pt ham or chicken stock
1 garlic clove, crushed
2tbls tomato purée
225g/8oz lean cooked ham, diced
boiled rice, to serve

- ***Serves 4-6***
- ***790cals/3190kjs per serving***

◀ 1 Cut the aubergines into 2cm/¾in dice. Put them into a colander and sprinkle them with salt. Leave to degorge for 30 minutes. Rinse with cold water and drain on absorbent paper.

2 Heat the oil in a large flameproof casserole over medium heat. Add the chicken, brown quickly on all sides and remove with a slotted spoon.

3 Lower the heat. Add the onion and cook until it is soft but not coloured, about 6-7 minutes. Mix in the paprika, cayenne pepper and aubergines. Cook for 1 minute, stirring.

4 Add the okra and sweetcorn. Pour in the stock and bring to the boil. Stir in the garlic and tomato purée. Add the chicken and ham. Cover the casserole and cook it over low heat for 25-30 minutes.

5 Take out the chicken pieces, remove the skin and discard, then take the meat from the bones and cut it into small cubes. Return the chicken meat to the casserole and reheat it briefly before serving with plain boiled rice.

Cook's tips

If you want your gumbo to have a thicker consistency, slice the okra before adding it to the mixture; this releases the sticky juices which act as thickening agents.

Bananas Foster

- **Preparation: 10 minutes**
- **Cooking: 10 minutes**

50g/2oz unsalted butter
75g/3oz soft brown sugar
¼tsp ground cinnamon
large pinch of grated nutmeg
125ml/4fl oz dark rum
4 firm, ripe bananas, peeled and halved lengthways
vanilla ice cream, to serve (optional)

- **Serves 4** (¶)(££)
- **310cals/1300kjs per serving**

1 Melt the butter in a large, shallow frying pan over low heat. Stir in the brown sugar, cinnamon and nutmeg. Add half the rum.

2 Add the halved bananas to the pan and cook until golden on all sides, turning carefully. Simmer for 2 minutes, then remove with a slotted spoon to a serving dish and keep warm.

3 Heat the remaining rum in a small saucepan. Ignite it and stir into the sugar mixture left in the pan. Pour slowly over the bananas. Serve warm, with ice cream, if wished.

SMOKE SCREEN

In the early days, the lack of refrigeration made smoking meat a necessity and the south to this day is famed for its hickory-smoked hams. Those from Smithfield, Virginia, are especially prized as the pigs are traditionally fed on peanuts which give their flesh a distinctive flavour and texture.

The small intestines of the pig were given to the slaves and played an important part in their diet. Called chitterlings, they were usually cooked in boiling water and spices until tender, then coarsely chopped, dipped in corn meal and deep fried in sizzling hot lard.

Cook's tips

It is not really worth buying ground nutmeg, as fresh nutmeg is easy to grate and the flavour is far superior. Whole nutmeg can be bought from most supermarkets and oriental food stores, and is easy to grate using the fine holes of a grater.

Down Mexico Way

A melting pot of exciting but simple ingredients and spicy flavours, the cooking of Mexico and Central America reflects its ancient Aztec origins and the influence of Europe

Almond jelly (page 38)

THE COOKING OF Mexico and the seven small countries of Central America – Belize, Guatemala, Honduras, El Salvador, Nicaragua, Costa Rica and Panama – remains firmly based on the native ingredients used before the Spanish Conquest. Corn (maize), chillies, tomatoes, beans and avocados were the staples; and in southern Mexico and Central America the use of rice, fish, coconut and exotic fruit shows the tropical and coastal influence. Potatoes and other root vegetables from South America were also used. Cocoa beans were used in trading and the Aztecs were known to have drunk a form of hot chocolate. The use of chocolate in savoury as well as sweet dishes is typical.

The European influence

Based on these ingredients, a very rich and varied style of cooking developed, utterly unlike anything known to invading Spanish conquistadors. The Spanish soon introduced their own foods such as wheat, olives, olive oil, wine, citrus fruits, cinnamon and nutmeg. More importantly, they also introduced cattle, which provided beef and dairy products, together with sheep, goats and the domestic pig. Olive oil and lard are the main fats used today. The new foods were enthusiastically incorporated into Mexican cuisine but even so, modern Mexican cooking sticks firmly to its origins, with corn, chillies, beans,

Cooking tortillas at a Mexican street market

tomatoes and avocados appearing at most meals in some form.

Cooking techniques

Certain cooking techniques that are fundamentally pre-Conquest have not changed either, although very few people still grind their chillies or corn with a stone mortar and pestle when a food processor can do the job in a fraction of the time.

The main meal of the day, *comida*, is eaten at midday or slightly later. It is a large meal which begins with soup, followed by a rice, pasta or tortilla dish called *sopa seca* (dry soup). Then follows a fish dish and the main course which would be a meat or poultry dish. Beans are sometimes served as a separate course, followed by a light dessert served with coffee or a herb tea such as camomile.

Tropical treats

Central American cooking has also been influenced by Africa, Britain and the Caribbean. In its subtropical climate, all the fruits and vegetables of the New World flourish, as do the introduced fruits and vegetables of Africa and Europe, such as okra, bananas and plantains. These fruits are so highly esteemed they are called *frutos de oro*, which translates as golden fruits.

Chayote soup

- ***Preparation: cooking chicken, then 20 minutes***
- ***Cooking: 35 minutes***

3 chayotes, peeled, seeded and sliced
salt and pepper
25g/1oz butter
1 onion, finely chopped
1 garlic clove, finely chopped
1tbls flour
1L/1¾pt chicken stock
225g/8oz cooked, shredded chicken breast
2tbls chopped fresh coriander, parsley or chives

- ***Serves 6-8*** 
- ***95cals/400kjs per serving***

1 Put the chayotes in a pan of salted water, bring to the boil, then simmer for about 20 minutes or until tender. Drain and reserve 500ml/18fl oz of the cooking liquid. Purée the vegetables in a food processor or blender, adding some of the reserved liquid if necessary. Transfer to a saucepan with the rest of the reserved liquid.

2 Heat the butter in a frying pan over medium-low heat and sauté the onion and garlic until soft. Stir in the flour and cook over a medium heat for 1 minute, stirring. Stir in enough of the chicken stock to make a smooth sauce. Add the sauce to the vegetable purée with the rest of the chicken stock, stirring to mix.

3 Season the soup with salt and pepper to taste and simmer for 2-3 minutes to blend the flavours. Add the chicken breast and simmer just long enough to heat the meat thoroughly. Serve in warmed bowls, garnished with the chopped herbs.

Cook's tips

Chayotes are at their best and most tender when small and pale; larger ones are best for stuffing. They can be stored in the refrigerator for up to a week.

SALSA CRUDA

This sauce is served at every meal. Finely chop 225g/8oz tomatoes, 2 or more fresh, hot, green chillies (seeds removed), 1 small onion and 1tbls fresh coriander. Combine and mix with salt and a pinch of sugar. Serve at room temperature.

Refried beans

No Mexican meal is complete without beans. Here they are cooked, then mashed and fried to make a smooth paste

- ***Preparation: 20 minutes***
- ***Cooking: 3¼ hours***

350g/12oz dried red kidney beans
2 onions, finely chopped
2 garlic cloves, chopped
2 small, dried hot red chillies
1 bay leaf, crumbled
lard or oil
salt
100g/4oz tomatoes, skinned and chopped
grated Cheddar cheese (optional)

- ***Serves 6-8***
- ***280cals/1175kjs per serving***

1 Wash and pick over the beans but do not soak them. Place in a large, heavy saucepan with enough cold water to cover by 5cm/2in. Add half the onions, garlic, the chillies and bay leaf. Cover, bring to the boil and boil vigorously for 15 minutes.

2 Reduce the heat and simmer gently, still covered, for about 30 minutes, adding hot water as necessary to keep the beans covered, until the beans begin to wrinkle. Add 1tbls lard or oil and continue to simmer for a further 15 minutes, adding more water if necessary to keep the beans submerged, until the beans are tender. Add salt to taste and simmer for 1 hour 15 minutes longer without adding any more water. There will be about 300ml/½pt liquid left when the beans are done.

3 Heat 1tbls lard or oil in a frying pan over medium-low heat and sauté the remaining onion and garlic for 10-12 minutes until soft. Add the tomatoes and cook for 5-6 minutes, stirring often, until the mixture is thick.

4 Add a ladleful of beans and their liquid to the mixture and cook over a medium heat, mashing with a potato masher to make a heavy paste. Repeat, adding 1tbls of lard or oil from time to time, until all the beans have formed a creamy paste. Sprinkle with grated cheese, if liked.

Ingredients guide

Clockwise from the top:
***Ancho** are perhaps the most commonly used. They are mild to hot, dried, wrinkled and deep red-brown.*
***Pasilla** are long, thin, brownish-black dried chillies with a pungent, piquant flavour.*
***Tortillas** are a thin flat bread, usually available in cans.*
***Masa harina** is a flour made from maize steeped and boiled in lime, fried and finely ground. It is used to make tortillas.*
***Jalepeño** are dark green and hot. Sold fresh or canned.*

Puebla-style turkey

This is Mexico's most festive and famous dish, known as *mole poblano*. The use of chocolate in a savoury dish is an older tradition than sweet chocolate dishes

- ***Preparation: 1½ hours***
- ***Cooking: 1½ hours***

about 8tbls oil
3.6kg/8lb turkey, cut into serving pieces
1 onion, chopped
2 garlic cloves, chopped
salt
For the sauce:
6 ancho chillies
6 mulato chillies
4 pasilla chillies
100g/4oz blanched almonds
50g/2oz roasted peanuts
4tbls sesame seeds
½tsp coriander seeds
¼tsp aniseed
2 whole cloves
1cm/½in piece of stick cinnamon, broken into small pieces
2 onions, roughly chopped
2 garlic cloves, chopped
450g/1lb tomatoes, skinned, seeded and chopped
75g/3oz seedless raisins
2 tortillas or 2 slices of toasted white bread, cut up into small pieces
40g/1½oz plain chocolate, broken up
salt
chilli flower, to garnish

- ***Serves 10*** 
- ***575cals/2415kjs per serving***

1 Heat 2tbls oil in a heavy frying pan over medium heat and sauté the turkey pieces, a few at a time, until they are lightly browned all over. Add more oil if the turkey starts sticking.

2 Transfer the turkey to a large, heavy casserole, add the onion, garlic, salt and enough water to cover. Bring to the boil, reduce the heat, cover and simmer for 1 hour or until the turkey is tender, rearranging the turkey pieces to ensure even cooking. Lift out the turkey pieces and reserve. Strain and reserve the stock. Rinse and dry the casserole.

3 Meanwhile, make the sauce: remove the stems and seeds from the three kinds of chillies. Tear them into pieces and put them into a bowl with 150ml/¼pt warm water. Let them stand for 30 minutes, turning the pieces from time to time.

4 Grind the nuts, half the sesame seeds, the coriander, aniseed, cloves and cinnamon in a blender or food processor. Transfer to a bowl and reserve.

5 In a blender or food processor combine the chillies, the water in which they have soaked, the onions, garlic, tomatoes, raisins and tortillas and blend to make a fairly thick paste.

6 Transfer the paste to a bowl and stir in the ground nuts and spices, mixing thoroughly. Measure the oil left in the frying pan after browning the turkey and make up the quantity to 4tbls. Put the pan over medium heat, add the paste mixture and cook, stirring, for 5 minutes.

7 Transfer the mixture to the casserole. Stir in 425ml/¾pt of the reserved turkey stock and the chocolate. Season and simmer, stirring until the chocolate is melted and the sauce is like heavy cream, adding more stock if necessary. Add the turkey pieces and simmer, covered, over low heat for 30 minutes, rearranging to ensure even cooking.

8 Arrange the turkey pieces in a large, deep serving dish and pour over the sauce. Toast the remaining sesame seeds and sprinkle on top. Garnish and serve.

Cook's tips

Chillies should be handled with care. Wear rubber gloves and avoid touching the face when preparing them.

Variations

If you can't find ancho, mulato and pasilla chillies, substitute 50-100g/2-4oz fresh green chillies and 2-4 dried red chillies, all deseeded.

MEXICAN RICE FOR 6-8

Soak 350g/12oz long-grain rice in hot water for 15 minutes. Drain, rinse and allow to dry thoroughly. Purée 1 large onion, 2 garlic cloves and 275g/10oz peeled tomatoes and reserve. Heat 3tbls oil in a frying pan and sauté the rice until golden. Transfer to a saucepan and add the purée, 1L/1¾pt chicken stock and seasoning to taste. Bring to the boil, then simmer, covered, over a very low heat for about 20 minutes. Stir in 50g/2oz defrosted frozen peas and cook until all the liquid is absorbed.

Tortillas

- **Preparation: 45 minutes**
- **Cooking: 40 minutes**

275g/10oz masa harina
1tsp salt (optional)

- **Makes about 13** 
- **85cals/355kjs each**

1 In a large bowl mix together the masa harina and salt, if using. Stir in about 300ml/½pt lukewarm water – enough to make a soft dough which should hold together nicely but not be stickily wet.

2 To form the tortillas, break off pieces of dough the size of small eggs. Roll between the palms to form a ball and place between two sheets of stretch wrap. Roll lightly to form a flat pancake 10cm/4in across. Stack them covered with a clean, damp tea-towel.

3 Heat a griddle or heavy frying pan over high heat. Gently place a tortilla on the griddle or frying pan. When the edges begin to curl, after about 2-3 minutes, turn the tortilla over and cook for 2-3 minutes longer. It should be slightly flecked with brown and may puff up in cooking.

4 Wrap the tortilla in a heavy napkin to keep it warm. As each one is cooked, stack it on top of the previous one, keeping them wrapped. Serve them in the napkin, keeping them covered so they stay warm and flexible.

VERSATILE TORTILLAS

Tortillas can be eaten in many ways besides as bread.
Tacos are stuffed and folded tortillas, fried in lard or oil and eaten with the fingers. Fillings include shredded chicken or minced beef in a chilli or tomato sauce, left-over Puebla-style turkey, spicy sausage with cubes of cheese, Refried beans – all topped with avocado, shredded lettuce and cheese.
Tostadas are crisply fried flat tortillas, with layers of lettuce, onion, refried beans, avocado, cheese and chilli sauce.
Enchiladas are rolled, filled and fried tortillas, covered with spicy sauce and cheese and baked.

Plan ahead

Tortillas can be made a day ahead and chilled, wrapped in their napkin and foil. To reheat, place the foil packet in the oven at 140C/275F/gas 1 for 15 minutes.

Corn and courgettes

- **Preparation: 10 minutes**
- **Cooking: 25 minutes**

25g/1oz butter
1 large onion, chopped
1 green pepper, seeded and roughly chopped
700g/1½lb courgettes, sliced
2 firm tomatoes, skinned and chopped
350g/12oz frozen sweetcorn kernels

- **Serves 6-8**
- **140cals/590kjs per serving**

1 Heat the oven to 180C/350F/gas 4. Melt the butter in a medium-sized casserole and sauté the chopped onion and pepper over a low heat until soft. Add the courgettes, stirring frequently until coated with butter.

2 Remove the casserole from the heat and add the tomatoes and sweetcorn. Season generously with salt and pepper and sprinkle with 4-6tbls water. Cover the pan and cook the corn and courgettes in the oven for 20 minutes or until tender. Serve immediately.

CHOCS AWAY

Ancient Mexicans used cocoa beans in a drink called *chocolatl*, considered an aphrodisiac. The beans were so highly prized, they were even used as currency. In the 16th century cocoa was taken back to Spain, where the secret of growing and preparing it was closely kept for a century.

Almond jelly

- ***Preparation: 45 minutes, plus chilling***
- ***Cooking: 10 minutes***

175g/6oz sugar
250ml/9fl oz boiling water
1/4-1/2tsp almond essence
15g/1/2oz powdered gelatine (1 sachet)
6 egg whites
pinch of salt
100g/4oz slivered almonds, toasted
For the custard:
600ml/1pt milk
6 egg yolks
3tbls sugar
1/2tsp vanilla essence
125ml/4fl oz double cream

- ***Serves 6*** 
- ***335cals/1410kjs per serving***

1 Put the sugar in a bowl, pour on the boiling water and stir until dissolved. Stir in the almond essence and cool.

2 Pour 25ml/1fl oz very hot water in a small bowl, sprinkle on the gelatine and leave to soften for 5 minutes. Stir to completely dissolve, standing the gelatine in a pan of hot water, if necessary. Stir this into the cooled almond syrup. Chill for 30 minutes or until it begins to thicken, then whisk for 2-3 minutes until frothy.

3 In a large bowl, beat the egg whites with a pinch of salt until stiff but not dry. Gently fold into the almond jelly.

4 Pour half the mixture into a dampened 1.5L/2½pt mould and sprinkle with half the toasted almonds. Carefully spoon the remaining mixture over the almonds. Refrigerate for 3-4 hours or until set.

5 To make the custard, heat the milk in a saucepan until almost boiling, then remove from the heat. Beat the egg yolks, sugar and vanilla for 5-6 minutes or until thick and pale yellow. Pour the hot milk onto the yolks, then return to the pan over medium heat. Cook, stirring constantly, until slightly thickened. Remove from the heat, cool and chill for 1-2 hours.

6 Whip the cream until soft peaks form, fold into the custard and chill.

7 To serve, unmould the pudding by dipping the mould into hot water for 15-20 seconds. Turn out onto a serving dish and sprinkle with the remaining almonds. Serve with the custard sauce.

Red snapper Veracruz-style

- ***Preparation: cooking potatoes, plus 25 minutes***
- ***Cooking: 45 minutes***

6 red snapper fillets weighing about 1kg/2 1/4lb
seasoned flour
100ml/3 1/2fl oz olive oil
1 onion, finely chopped
1 garlic clove, finely chopped
400g/14oz canned chopped tomatoes
3tbls capers
large pinch of ground cinnamon
large pinch of ground cloves
3 canned jalepeño peppers or fresh green chillies, seeded and cut into strips
2tbls lemon or lime juice
1/2tsp sugar
salt and pepper
12 small stuffed green olives, rinsed
butter, for frying
3 slices firm white bread, cut into triangles
12 new potatoes, freshly cooked, peeled and halved

- ***Serves 6*** 
- ***370cals/1555kjs per serving***

1 Dust the fish fillets generously with seasoned flour, shaking to remove the excess. Heat half the oil in a frying pan over medium-low heat and fry the fillets for about 5 minutes until they are golden on both sides. Keep warm.

2 Sauté the onion and garlic in the remaining oil until soft. Add the tomatoes, capers, cinnamon, cloves and jalepeño peppers and simmer for 5 minutes to blend the flavours. Add the lemon or lime juice, sugar, and salt and pepper to taste. Simmer for 2 minutes more. Add the fish and olives and heat through. Remove from the heat and keep warm.

3 Heat the butter in a frying pan over medium low heat and fry the bread until golden. Drain on absorbent paper.

4 Arrange the fish on a heated serving dish surrounded by the potatoes and fried bread. Serve at once.

Variations

If you can't get hold of red snapper, red sea bream or any firm, non-oily white fish will make a good substitute.

Way Out West

The cookery of western America mirrors its history, beginning with the original American Indians and encompassing the cuisines of successive waves of colonists and immigrants

Hopi corn and squash casserole (page 41)

THE AMERICAN WEST is a vast area, including the cattle-ranching and desert Southwest states of Arizona, New Mexico, Texas and Oklahoma; the West Coast, mainly the state of California, lying along the blue Pacific; and the Midwest, stretching from the corn-growing plains of Missouri, Kansas and more northerly states, to the Rocky Mountains bisecting Colorado and the plateau with Utah and Nevada.

Indian vegetables

The original American Indians grew sweetcorn, beans and squash, which they taught the early settlers to use. Until 1848 California was governed by Spain, and later Mexico. The Spanish brought their own style of cooking, plus the chilli pepper from South America, and it became the hallmark of the cooking of Southern California and all states bordering on Mexico. Spanish settlers also brought their native peaches, citrus fruits and tomatoes.

Railroads

Some Chinese settlers came to the port of San Francisco to build the

trans-continental railroad, bringing bean sprouts, bamboo shoots and other Chinese foods. The completed railroad brought Italian immigrants, along with their own rich cooking traditions. They also turned much of the Napa Valley into flourishing vineyards, making California today a big wine exporter. In the Midwest many immigrants came from Central and Eastern Europe and Scandinavia, all contributing different cooking.

Local produce

As everywhere, the cuisine is a product of the locally available foodstuffs. In the early days only corn (maize, or Indian corn) was grown on the vast Midwestern plains. It was the staple food, used to make corn bread, corn mush, corn casseroles and hominy grits. Lacking refrigeration, settlers dried, salted and pickled the vegetables they grew; cabbage was very efficiently preserved as sauerkraut, a German creation. But the staple pioneer diet, especially in the Midwest, was game. Bear, deer, elk, and originally large herds of buffalo, were as plentiful as the smaller game – wild turkey, squirrels, opossum and rabbits.

Later settlers in Colorado and Nevada began to raise cattle and sheep, and the cookery remained heavily meat-oriented. In the corn states farmers raised an abundance of hogs, and their wives became famous for home-cured hams.

Shopping at Fisherman's Wharf, San Francisco

SEAFOOD SPECIALS

South-eastern Texas borders the Gulf of Mexico, which provides a wealth of prawns and oysters, red snapper and pompano, plus abundant crabs. The preparation of seafood shows the deep-south influence of neighbouring Louisiana, with dishes such as crab gumbo and baked-in-parchment pompano, often accompanied by corn bread.

Seafood also features largely in the cooking of the West Coast. The magnificent Pacific salmon are served baked, barbecued, stuffed, fried, pickled and even baked in a soufflé. Crabs are plentiful, the king-sized Dungeness yields six times more than its Atlantic cousin.

Frisco crab

● ***Preparation: 20 minutes***

450g/1lb crabmeat, cooked
lettuce or trimmed spinach
200ml/7fl oz mayonnaise
75ml/3fl oz thick cream
50ml/2fl oz chilli sauce
25g/1oz green pepper, seeded and chopped
2tbls chopped spring onion
2tbls mustard pickle or piccalilli
lemon juice
salt
dash of Worcestershire sauce
4 eggs, hard-boiled and sliced
4 tomatoes, sliced
parsley or olives

● ***Serves 4*** 

● ***640cals/2690kjs per serving***

1 Pick over the crab removing any pieces of shell. Arrange the crab on a bed of lettuce or spinach leaves on a large platter or on salad plates.

2 Combine the next nine ingredients in a bowl for the sauce and season to taste. Spoon the sauce over the crab, and garnish with slices of hard-boiled eggs, sliced tomatoes and parsley or olives.

THE CORNBELT

The American cornbelt consists of the states located on flat fertile plains in the heartland of the American continent: Kansas, Missouri, Iowa, Nebraska, North and South Dakota. Corn means maize (sweetcorn) in the United States; but the so-called corn states also produced vast acres of what we call corn: wheat.

Californian fish stew

(Cioppino)

This spectacular stew of Italian inspiration (pronounced chopeeno) is adapted to what is available; there is no right way to make it. White wine is used in the dish but red wine is often served with it

- **Preparation: 1 hour**
- **Cooking: 45 minutes**

450g/1lb sea bass or any other firm flaky fish, cleaned, boned and cut into 5cm/2in cubes
450g/1lb prawns, shelled
450g/1lb crab meat, cooked
12 raw clams
6 raw oysters
1-2 small lobsters, cooked and cut in their shells into 6 pieces
3tbls olive oil
50g/2oz butter
2 onions, coarsely chopped
1 large green pepper, seeded and diced
2-3 large garlic cloves, finely chopped
450g/1lb fresh plum tomatoes, blanched, peeled and coarsely chopped or canned plum tomatoes, drained and coarsely chopped
3tbls tomato purée
1 bay leaf
½tsp dried thyme
6 peppercorns
salt
450ml/16fl oz good quality dry white wine
2tbls finely chopped parsley

- **Serves 6**
- **535cals/2245kjs per serving**

1 Heat the olive oil and butter in a large frying pan over high heat. Add the onions, green pepper, and garlic and cook, stirring frequently, until the onions just begin to brown. Add the tomatoes, tomato purée, bay leaf, thyme, peppercorns, salt and 300ml/½pt of the wine. Cook, covered, for 30 minutes over very low heat, then season to taste.

2 Pour a thin layer of the vegetable sauce into the bottom of a very large flameproof casserole. Place the fish, prawns and crabmeat in the sauce and pour over the remaining sauce. Add the remaining wine. Place the lobster and then the clams and oysters on top. Cover the pot and cook over medium high heat for about 12 minutes, or until the fish is cooked and the clams or oysters have just opened. Do not over-cook the fish stew as this will make the texture of the seafood rubbery and spoil the dish.

3 Serve from the casserole or from a very large bowl, arranging the clams and oysters on top and sprinkling with parsley. Serve the stew in warmed soup bowls and provide bowls for emptied shells.

Variations

Use fresh fish available. Substitute mussels for clams and langouste for lobster.

Hopi corn and squash casserole

- **Preparation: 25 minutes**
- **Cooking: 25 minutes**

8 large corn on the cob or 450g/1lb frozen corn
1tbls vegetable oil
1kg/2¼lb courgettes or other soft-skinned summer squash, cubed
300g/11oz onion, coarsely chopped
2 fresh green chillies, chopped
1tsp finely chopped garlic
500g/18oz tomatoes, blanched, peeled and coarsely chopped
500g/18oz fresh, frozen or canned lima beans
250g/9oz green beans, trimmed and cut into 3
salt and freshly ground pepper

- **Serves 8**
- **335cals/1405kjs per serving**

1 Slice the kernels from the corn with a sharp knife and reserve.

2 In a large saucepan heat the oil. Add the courgettes or squash, onion, chillies, garlic, tomatoes, lima beans and green beans. Cook over a low heat, stirring until the onions are translucent.

3 Add 250ml/9fl oz water, salt and pepper, bring to the boil and then simmer, covered, until the beans are almost done, about 10-12 minutes.

4 Add the corn and cook about 5 minutes (or longer if using frozen). Season to taste and serve hot.

Beef stew with olives

• Preparation: 25 minutes

• Cooking: 1 hour 35 minutes

75ml/3fl oz olive oil
1 large onion, coarsely chopped
1-2 garlic cloves, finely minced
1.5kg/3lb 5oz lean stewing beef, cubed
seasoned flour
200ml/7fl oz red wine
24 large green olives
salt and pepper

• Serves 6 

• 580cals/2435kjs per serving

1 Heat half the oil in a large flameproof casserole over high heat. Cook the onions and garlic, stirring frequently, until they begin to brown. Remove the onions and garlic and reserve.

2 Toss the cubed beef in seasoned flour to coat thoroughly. Shake off excess flour. Heat the remaining olive oil in the casserole. Fry the beef in batches, turning the cubes until they are browned on all sides. Return the onions and garlic to the pot and mix well. Scrape the bottom to release the juices.

3 Add the red wine and enough water to cover the meat, and bring to the boil. Cover and cook over low heat, stirring occasionally, for about 1½ hours until the meat is tender. About 5 minutes before you are ready to serve, add the olives, stirring to distribute them. Taste and season before serving with rice or mashed potatoes and a selection of varied vegetables.

Thousand island dressing

California is famous for its salads and its salad dressings such as this one

• Preparation: 10 minutes

225ml/8fl oz mayonnaise
1tbls tomato purée
1-2tbls tarragon vinegar
1 garlic clove, finely chopped
pinch of cayenne pepper
2tbls chopped celery
2tbls chopped dill pickle
2tsp Worcestershire sauce

• Makes about 300ml/½pt 

• 1820cals/7645kjs total

1 Mix the mayonnaise with the rest of the ingredients, correct the seasoning and serve immediately or cover and refrigerate.

Cook's tips

Serve with sliced avocados, tomatoes, cucumber, hard-boiled egg, green salad or green beans.

TEX-MEX CUISINE

The Southwest and West states all border on or are very close to Mexico, which has given rise to a unique cuisine known as Tex-Mex; a blend of Southwestern taste with Mexican cooking. Every fast food restaurant has *tacos* and *enchiladas* – cornmeal pancakes with spicy fillings – *guacamole* – spicy avocado dip; *huevos rancheros* (Ranch style eggs); *salsa verde* – green chilli sauce; and *buñuelos* – deep-fried syrup-coated dough puffs.

But by far the most famous Tex-Mex dish is *chilli con carne* which neatly sums up the foreign influences on the region's cooking. Recipes vary, but all contain the tomatoes and chillies introduced by the Spanish, the native American red kidney bean and Britain's beef.

Land of The Pilgrim Fathers

A whole new way of living awaited the New England settlers – and a variety of wild food destined to form the basis of traditional American cooking

A hearty old-fashioned American breakfast includes eggs, bacon, fried scrapple (page 45) and coffee

WHEN THE PILGRIM Fathers arrived in America in the 17th century, their stocks of food had gone, but a land of plenty awaited them, packed with game, fruit and vegetables they had never seen before. Hunger is a good teacher and it was not long before the New England colonists were busily adapting Indian staples to their own way of life and eating.

Corn pone and clam chowder

Bread was, as always, the prime requirement and the settlers were soon cooking cornmeal with water to make corn mush, a quick and filling porridge. Cooked over a slow fire, this could make corn dough, which in turn was baked into corn pone. Seafood was plentiful: clams were dug up and made into the now famous clam chowder. Fishing boats were soon out in the Atlantic searching for cod.

Talking turkey

Turkeys were a bonus and were soon a special part of the settlers' life, later to be eaten at Thanksgiving with cranberry sauce, oyster stuffing and corn bread. The long hard New England winters led to the creation of the boiled dinner, a combination of boiled beef and traditional pot vegetables. Pumpkins, winter squash and sweet corn were new vegetables for the English; wild cranberries and blueberries were gathered and used for pies and puddings. English apples went native and soon American apple pie was on the menu of many American families.

Pennsylvania pickles

Moving further south, the colonists were followed by Dutch settlers who brought different cooking traditions with them together with stocks of wheat, rye and barley and all the home baking of Holland – waffles, pancakes, doughnuts and cookies. Then came German farmers, nicknamed Pennsylvania Dutch, a corruption of Deutsch their cooking was hearty economical country fare based on pork products, spicy stews, smoked meats and sausages. They were served with superb pickles and preserves such as spiced pears, apple butter and watermelon rind as well as ketchups based on local tomatoes. It is no accident that the greatest professional picklemaker of all time came from Pennsylvania – Mr 57 varieties, Henry Heinz.

Farm stand selling squash

Philadelphia pepperpot

Legend has it that this dish was created during the War of Independence at Valley Forge when General George Washington demanded his cook create a dish to feed his troops. 'I have only tripe and peppercorns' complained the cook, but he then produced this tasty innovation

- ***Preparation: 40 minutes***
- ***Cooking: 4 hours***

400g/14oz fresh honeycomb tripe, parboiled
1kg/2¼lb knuckle of veal
1 small bunch parsley
16 black peppercorns, coarsely crushed
12 whole cloves
¼tsp each fresh basil, savory, thyme and marjoram
50g/2oz butter
3 green peppers, coarsely chopped
3 large onions, coarsely chopped
2 large beetroots, diced
2 large potatoes, diced
2tsp salt
50g/2oz long-grain rice

- ***Serves 8*** 
- ***275cals/1155kjs per serving***

1 Wash the tripe very thoroughly and cut it into small dice. Put the veal knuckle in a large, heavy casserole, add the tripe pieces and cover with water. Bring to the boil and cook for 15 minutes, skimming frequently.

2 Cover the pot, lower the heat and simmer gently for about 2 hours.

3 Tie the herbs and spices in a cheesecloth and add it to the pan. Cover the pot again and continue to cook for a further 1 hour.

4 Melt the butter in a large frying pan and sauté the vegetables until the onions and potatoes are nicely browned. Add the vegetables, salt and rice to the pot. Taste the liquid and, if it is well seasoned, remove the cheesecloth bag (otherwise leave it in). Cover and simmer an additional 30 minutes.

5 Remove the veal knuckle, reserving it for another meal. Discard the herbs in the cheesecloth bag. Allow the soup to cool and skim the fat from the surface. Reheat when ready to use.

Cook's tips

It is better to prepare this dish the day before it is to be eaten. This allows the fat to solidify on the surface when it is easily removed. The flavour is excellent when reheated.

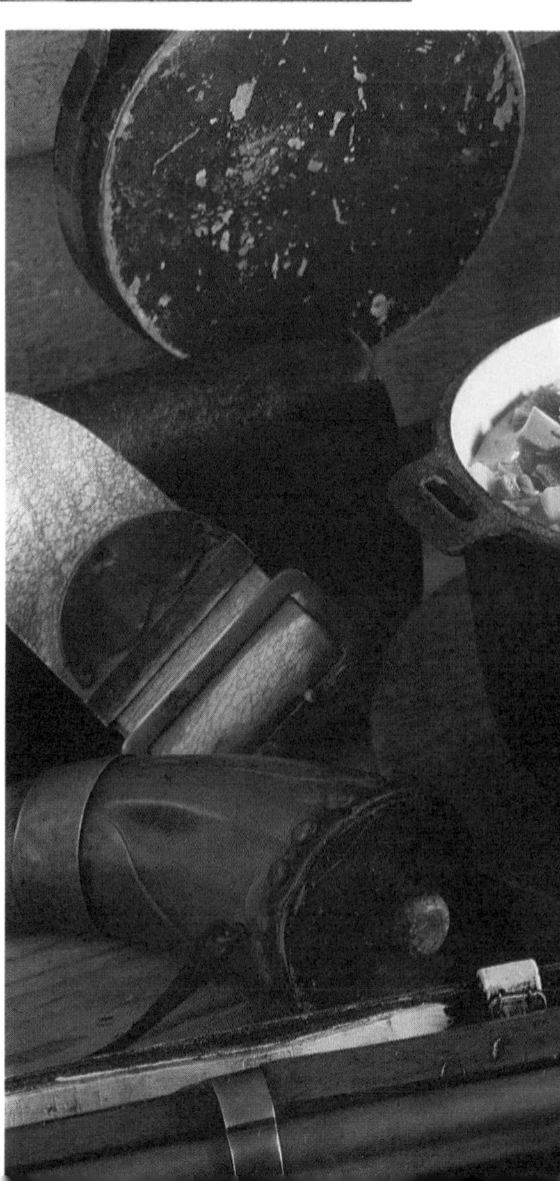

Scrapple

Sliced and fried, scrapple is a breakfast favourite among traditional farmers of the Pennsylvania Dutch country. It was a logical by-product of butchering days, when a good broth was always made

- ***Preparation: 10 minutes***
- ***Cooking: 1¼ hours, plus cooling time***

1½L/2½pt beef broth or water
225g/8oz cooked lean pork, diced small
150g/5oz cornmeal
1tsp salt
¼tsp freshly ground black pepper
¼tsp sage (optional)
butter for frying
fried eggs and bacon to serve

- ***Serves 8*** 
- ***425cals/1785kjs per serving***

1 Put the broth or water in a large, heavy bottomed casserole, add the pork and bring to a simmer for 10 minutes, then bring up to a rolling boil.

2 Slowly sprinkle in the cornmeal, whisking vigorously all the while to avoid lumps. Continue stirring until the mixture resembles thick porridge.

3 Add the seasonings and cook over low heat for 1 hour, stirring from time to time to stop the mixture sticking to the bottom of the pan.

4 Pour the mixture into a greased 1.7L/3pt tin and set aside to cool. When cold, turn out and slice the scrapple. Fry the slices in butter or lard and serve them hot with fried eggs and bacon for a hearty breakfast.

BEANS IN THE POT

The famous Boston baked beans were originally cooked in a large pot in a pit lined with hot stones so that they could be left to cook all through Saturday and be eaten when the Puritan Sabbath began. Molasses and salt pork gave them their distinctive rich flavour.

Whether or not to add mustard and an onion is a matter of heated debate!

New England boiled dinner

- ***Preparation: 15 minutes***
- ***Cooking: 4 hours***

2kg/4½lb salt beef
6 medium-sized potatoes, peeled and halved
1 large swede, peeled and cut into 5cm/2in dice
10 small pickling onions, peeled
6 medium-sized carrots, cut into thirds
1 small head cabbage, cut into sixths

- ***Serves 8*** 
- ***605cals/2540kjs per serving***

1 In a very large stewing pot or casserole cover the salt beef with about 2.5cm/1in of water and bring to the boil.

Skim off the foam, then simmer for 3½ hours, or until the beef is tender when pierced with a fork. Add water as needed to keep the beef covered.

2 About 30 minutes before the beef is done, add the prepared vegetables one type at a time, waiting until the liquid resumes bubbling before adding the next vegetable.

3 Drain the beef and vegetables and place the carved beef in the centre of a large platter. Surround the beef with the vegetables before serving.

Cook's tips

If it is not convenient to cook the meat on the top of the stove, put it in the oven at 160C/325F/gas 3. Remove and finish off step 2 on top of the stove. Use left over stock for gravy or soup.

ALL SQUASHED UP

Seeds for typical English vegetables were soon brought to New England, but in the meantime, the colonists experimented with the acorn and butternut squash, sliced in half, baked and served as a sweet or savoury. Pumpkin pie (page 48) was soon an all-American favourite, eaten and enjoyed at Thanksgiving and on the 4th July.

Chicken à la King

- **Preparation: 10 minutes, plus cooking the chicken**
- **Cooking: 15 minutes**

75g/3oz butter
150g/5oz mushrooms, thinly sliced
2tbls flour
salt and freshly ground white pepper
250ml/9fl oz thin cream
450g/1lb cooked chicken (preferably breast meat), diced into small pieces
3 egg yolks
1tsp very finely chopped onion
2tbls lemon juice
2tbls dry sherry
1 green or red pepper, cut into dice
buttered noodles or hot buttered toast to serve

- **Serves 4** 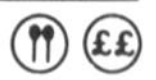
- **700cals/2940kjs per serving**

1 Melt 25g/1oz butter in a large saucepan over medium-low heat. Sauté the mushrooms for 5 minutes, stirring frequently. Stir in the flour and season well with salt and pepper.

2 Stir in the cream and cook, uncovered, over medium heat until the mixture thickens, stirring frequently. Pour the sauce into the top pan of a double boiler over gently simmering water. Add the chicken and heat it through thoroughly.

3 Meanwhile, beat the remaining butter in a bowl with a wooden spoon, then beat in the egg yolks. Add the onion, lemon juice and sherry. Add this slowly to the chicken mixture, stirring all the time. Continue to cook a few more moments, add the diced pepper and adjust the seasonings. Serve immediately on buttered noodles or with buttered toast.

FIT FOR A KING

Boneless cooked chicken gently reheated in a rich cream sauce, Chicken à la King (page 46), was named not for George III but for E. Clarke King III; it was a special treat concocted for him by the chef at the Brighton Beach hotel near Coney Island. Waldorf salad was another hotel creation: the maître d'hôtel at the Waldorf was the man who invented this crunchy combination of apples, celery, walnuts and mayonnaise.

Shoo-fly pie

The sticky molasses filling of this pie has a great attraction for flies

- ***Preparation: 45 minutes***
- ***Cooking: 30 minutes***

125ml/4fl oz molasses
125g/4oz soft light brown sugar
½tsp bicarbonate of soda
2 medium-sized eggs, beaten
175g/6oz flour
¼tsp salt
¼tsp ginger
¼tsp nutmeg
½tsp cinnamon
25g/1oz cold butter, diced
125g/4oz raisins
ice cream or thick cream, whipped, to serve

For the crust:
150g/5oz flour, sifted
½tsp salt
125g/4oz lard or vegetable shortening
3tbls (or slightly more) iced water

- ***Serves 8*** 
- ***575cals/2415kjs per serving***

NEW YORK, NEW YORK

During the 19th century, millions of immigrants from all over the world passed through Ellis Island to New York. As a result, you can find food from every cuisine in the world. Chinatown offers food from every part of China and in Manhattan's Little Italy there is Sicilian food to rival the best in Palermo. In Lindy's famous Times Square restaurant, you can eat their New York cheesecake, an American variant of an old Eastern European dessert.

1 To make the crust, mix the flour and salt in a large bowl. Cut in the lard with 2 knives until the mixture resembles oatmeal.

2 Sprinkle the mixture with iced water while stirring with a fork until the mixture forms a ball.

3 Wrap in stretch wrap and allow to rest in cool place for 20 minutes. Flatten it, then roll on a floured board to a circle about 3mm/⅛in thick. Fit into a 25cm/10in pie plate or flan tin. Heat the oven to 230C/450F/gas 8.

4 Stir 125ml/4fl oz hot water into the molasses and brown sugar in a bowl. Add the bicarbonate of soda and eggs and stir well.

5 Mix the flour, salt and spices in another bowl. Work in the butter until the mixture resembles fine breadcrumbs.

6 Sprinkle the raisins over the pastry case, then add the molasses mixture, alternately with the crumb mixture, finishing with a layer of crumbs.

7 Bake for 10 minutes, reduce the heat to 180C/350F/gas 4 and bake for 20 minutes, or until the top feels fairly firm. Serve warm or at room temperature with ice cream or thick cream.

Pumpkin pie

Few Americans would dream of celebrating Thanksgiving without pumpkin pie for dessert. If using fresh pumpkin, buy 1kg/2¼lb and stew it over low heat until virtually all of the liquid has evaporated

- ***Preparation: 10 minutes, plus making pastry and cooking pumpkin***
- ***Cooking: 40 minutes***

2 large eggs
100g/4oz sugar
1tbls molasses
¼tsp ground ginger
½tsp freshly grated nutmeg
1tsp ground cinnamon
pinch of ground cloves
½tsp salt
450g/1lb canned or stewed pumpkin
375ml/13fl oz milk
25cm/10in unbaked pastry case (see Shoo-fly pie, page 47)
275ml/10fl oz double or whipping cream, whipped

- ***Serves 8***
- ***425cals/1785kjs per serving***

1 Prepare the pumpkin by cutting in portions. Remove the skin and seeds. Cut the slices of flesh into chunks.

2 Heat the oven to 200C/400F/gas 6. Combine the eggs with the sugar, molasses, spices and salt; blend well. Add the pumpkin and milk, mix well and check the spices. Pour into the pastry case.

3 Bake in the centre of the oven for 40 minutes or until set. Serve warm or cold with sweetened whipped cream.

ALIEN CORN

The settlers soon learnt that the dough made from corn mush was excellent food for travelling. Fried on a griddle, it made a hard but nourishing bread called 'johnnycake' which was probably a corrupt form of 'journeycake', bread for taking on an expedition. They found that corn mush tasted all the better for the addition of maple syrup, a favourite sweetener, and cooked with sugar, eggs and cream, it made Indian pudding.

South American Way

The cooking of Spain and Portugal blends with the flavours of Africa and native Indian produce to create the rich and varied flavours of South American fare

Chilean fish soup (page 50)

THE CORDILLERAS OF Chile, the pampas of Argentina, the brilliant sun and icy shade of the Andes, the steamy jungles of Brazil – was there ever a continent so varied? The climate varies equally and this variety is echoed in the different dishes of South America.

Steaks large and juicy

Argentina is great cattle-ranching country and the early Spanish settlers were quick to ship livestock over from Europe – and in more recent years to send it back frozen in cold storage. The traditional Argentine barbecue scorns the individual steak and roasts a whole steer on long, strong, massive skewers. In Brazil, this barbecued beef is served traditionally with toasted cassava meal and lashings of hot chilli sauce.

Indian specials

By the time the first settlers arrived, the local Indians had begun to cultivate native vegetables such as pumpkin, squash, sweet corn, avocados and tomatoes. Cassava, the starchy tuber also known as manioc, became very popular with the African slaves taken to Brazil by the Portuguese but unlike the Indians they prepared it with the nutty palm oil brought from their home country. One of South America's great gifts to Europe is the potato, first cultivated by the Incas in Peru and introduced to Britain by Sir Walter Raleigh. Vegetables in the high Andes are normally served crisply tender – this is not the influence of *nouvelle cuisine* but the effect of cooking at high altitudes where water boils at a lower temperature.

Not so chilly

Hot and spicy is how they like their food in South America; hot pepper sauce, often prepared with fresh lime juice (see below), appears on every table and most stews and casseroles will contain a substantial amount of chillies as well as garlic, cayenne pepper, sweet paprika and coconut. The dish *Causa a la chiclayana* (Potatoes with fish and vegetables, page 53) illustrates a typical blending of flavours: sweet potatoes, ordinary potatoes, cassava root and green bananas are blended with chillies and vinegar to form a rich base for white fish fillets.

Boiling points

It's not unusual for newcomers to Ecuador to suffer from altitude problems. Quito, the capital, lies over 3000m. above sea level and enjoys great extremes of sun and shade. It is impossible to boil at full heat, so – making a virtue of necessity – the local cooking specialises in the long, slow simmering of casseroles. Nothing is ever as hot in temperature as it could be; this is where chilli sauce comes in extra handy and is used extensively.

All the colours of a South American market.

Chilean fish soup

- ***Preparation: 25 minutes***
- ***Cooking: 25 minutes***

1.5kg/3¼lb cod, bass, or other firm-fleshed non-oily white fish fillets cut into 6 slices
3tbls oil
1tsp paprika
2 onions, finely chopped
1 green or red pepper, seeded and finely chopped
2 carrots, thinly sliced
1kg/2¼lb potatoes, peeled and quartered
1 bay leaf
¼tsp dried oregano
3 parsley sprigs
1.5L/2½pt fish stock or water
2tbls chopped fresh coriander
crusty bread to serve

- ***Serves 6*** 
- ***505cals/2120kjs per serving***

1 Heat the oil in a large flameproof casserole over medium heat and sauté the onions until soft but not browned, add the paprika.

2 Add the pepper, carrots and potatoes, bay leaf, oregano and parsley sprigs. Stir, and cook for 2-3 minutes. Pour in the fish stock or water, cover and simmer until the potatoes and carrots are almost tender, about 15 minutes.

3 Add the fish, cover and simmer until the fish is tender, about 10 minutes. Remove and discard the bay leaf and parsley. Ladle the soup into a warmed tureen, sprinkle with the coriander and serve immediately with crusty bread.

Hot pepper and lime sauce

Brazilians are fond of hot food but they recognize that tastes differ, so a small bowl of hot pepper sauce is always on the table to be taken at the diner's discretion. This sauce is traditionally served with black bean stew, but is also used to accompany any meat, poultry or fish

- ***Preparation: 10 minutes***

3-4 hot red or green chillies, seeded and coarsely chopped
1 onion, chopped
1 clove garlic, chopped
salt
100ml/3½fl oz lime or lemon juice

- ***Makes 175ml/6fl oz*** 
- ***40cals/170kjs total***

1 Combine all the ingredients in a blender or food processor and blend to a purée; or crush in a mortar with a pestle, adding the lime or lemon juice bit by bit. Pour into a small bowl to serve.

Cook's tips

Store extra sauce, covered, in the refrigerator; use to add a dash of fire to other dishes.

FISH HARVEST

The cold Humboldt current sweeps up the coast of Chile and presents inshore fishermen with a harvest of fish which would be luxury items elsewhere: there are oysters, giant crabs, abalone and a unique white-fleshed fish called *congrio*, not to be confused with the Spanish *congrio* which is, of course, an eel.

Rice flour pudding

Moulded savoury puddings made of rice flour or cornflour are popular in Brazil as accompaniments to Afro-Brazilian dishes like Chicken in shrimp and nut sauce. Served at room temperature, or very lightly chilled, their bland flavour contrasts well with the exotic flavours of Bahian food

- ***Preparation: 5 minutes***
- ***Cooking: 5 minutes, plus 20 minutes to stand***

150g/5oz rice flour
1tsp salt
350ml/12fl oz thin coconut milk (see page 140)
225ml/8fl oz thick coconut milk (see page 140)

- ***Serves 8-10*** 
- ***155cals/650kjs per serving***

1 Combine all the ingredients in a saucepan, stirring to mix well. Cook over low heat, stirring constantly with a wooden spoon until the mixture is smooth and thick, about 5 minutes.

2 Turn the mixture into a buttered bowl and let it stand for 20 minutes. Turn out on to a serving dish and serve at room temperature. Sometimes it is served very slightly chilled.

Cook's tips

You can use creamed coconut and hot water to make the coconut milk.

Chicken in shrimp and nut sauce

This is one of the greatest dishes of the Bahia and it is magnificent for a party. The hard work once involved in making the dish has disappeared with the use of electric food processors, coffee mills, or blenders.

- ***Preparation: 45 minutes***
- ***Cooking: 1¼ hours***

3tbls olive oil
2 onions, finely chopped
4 spring onions, chopped, using white and green parts
2 large garlic cloves, chopped
500g/18oz tomatoes, blanched, skinned and chopped
1 or 2 fresh hot red or green chillies, seeded and chopped
salt and freshly ground black pepper
3tbls lime or lemon juice
4tbls fresh coriander leaves or flat-leafed parsley, chopped
2 × 1kg/2¼lb chickens, quartered
chicken stock, if necessary
225g/8oz peanuts, cashews or almonds, finely ground
225g/8oz dried shrimps, finely ground
675ml/24fl oz thin coconut milk (see page 140)
225ml/8fl oz thick coconut milk (see page 140)
1tbls rice flour
4tbls palm oil
Rice flour pudding to serve (see above)

- ***Serves 8-10*** 
- ***1045cals/4390kjs per serving***

1 Heat the oil in a large frying pan over medium heat and sauté the onions, spring onions, garlic, tomatoes and chilli peppers for 5 minutes. Season with salt and pepper, stir in the lime or lemon juice and the coriander or parsley.

2 Add the chicken pieces, cover and cook until the pieces are tender, 35-40 minutes. If the mixture seems dry, add 125ml/4fl oz chicken stock.

3 Lift out the chicken pieces and let them cool. Skin and bone the chicken and chop the meat coarsely. Put the mixture remaining in the frying pan through a sieve, pressing down hard on the vegetables to extract all the juices. Discard the solids and reserve the liquid.

4 In a heavy casserole combine the ground nuts, ground shrimps, the thin coconut milk and the reserved liquid. Stir on a high heat, reduce the heat and simmer for 15 minutes.

5 Stir in the thick coconut milk. Mix the rice flour to a smooth paste with a ▶

little cold water and stir it into the casserole. Cook, uncovered, over low heat, stirring frequently, until the mixture has the consistency of a thick béchamel sauce.

6 Add the chicken and the palm oil and cook just long enough to heat through. Transfer to a heated serving dish and serve with Rice flour pudding.

Stuffed skirt steak

The Spanish name of this satisfying dish means 'kill hunger', which indeed it does and without bankruptcy since skirt steak is modestly priced. It can be eaten hot as a main course and, in smaller amounts, as a first course, or cold with salads and for picnic fare

- ***Preparation: 30 minutes***
- ***Cooking: 2 hours***

550g/1¼lb skirt steak
1tsp dried oregano
2 large garlic cloves, crushed
salt and freshly ground black pepper
cayenne pepper (optional)
100g/4oz fresh spinach leaves, trimmed
1 carrot, thinly sliced
1 hard-boiled egg, thinly sliced
1 onion, thinly sliced and divided into rings
2L/3½pt beef stock
Hot pepper and lime sauce (page 50), to serve

- ***Serves 4*** 
- ***305cals/1280kjs per serving***

1 If the steak is thick, ask your butcher to butterfly it by cutting the steak horizontally to within about 1.5cm/½in from one of the longer sides. It is opened and flattened lightly with a cleaver. Alternatively do this yourself – if the steak is a thin one, do not bother with this step.

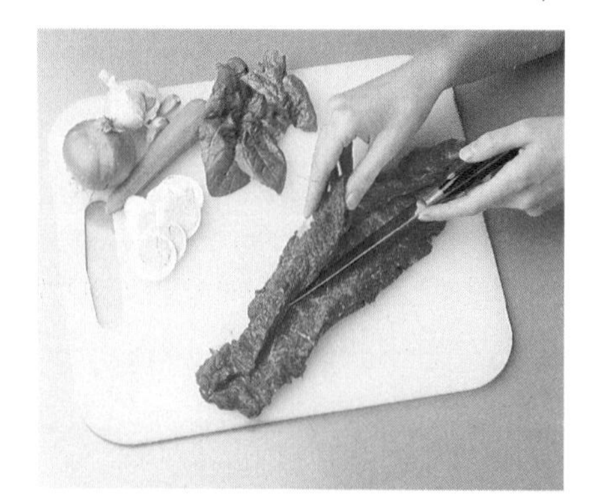

2 Mix the oregano and garlic together and spread them over the steak; then season with salt, pepper and cayenne if you like. Arrange the spinach leaves over the steak, then top with the carrot and egg. Scatter the onion rings over the top.

3 Roll up the steak like a Swiss roll and tie with string at 2.5cm/1in intervals. Put the steak into an oval casserole into which it will fit snugly.

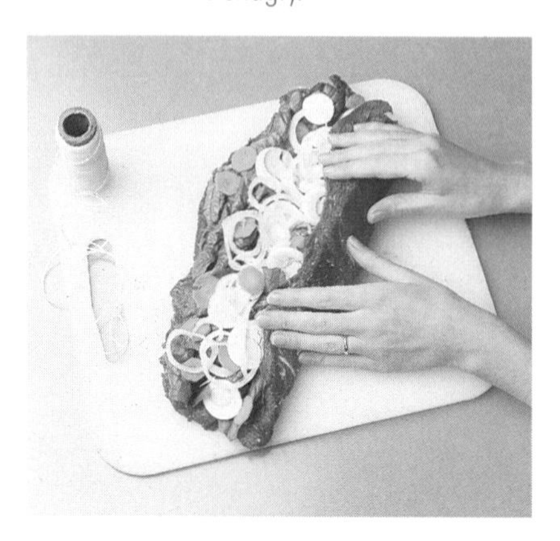

4 Pour in the beef stock which should cover the steak. Bring the liquid to the boil, skim off any froth that rises, reduce the heat and simmer gently, covered, for 1½-2 hours, or until the steak is tender when tested.

5 Lift out the steak, remove the string and cut the meat into 1.5cm/½in slices. Arrange it on a warmed platter and moisten with a little of the stock; then serve with hot pepper sauce, potatoes or rice and a green vegetable.

Potatoes with fish and vegetables

- **Preparation: 45 minutes**
- **Cooking: 1½ hours**

1 small onion, finely chopped
100ml/3½fl oz lemon juice
large pinch cayenne pepper
salt and freshly ground black pepper
1.5kg/3¼lb potatoes, peeled and halved
400ml/14fl oz olive oil
450g/1lb sweet potatoes, cut into 6 slices
450g/1lb cassava root, peeled and cut into 6 slices (optional)
3 green bananas
2 corn on the cob
1kg/2¼lb fillets white fish, cut into 5cm/2in pieces
seasoned flour
3 large fresh hot red or green chillies (about 10cm/4in long)
3 onions, thinly sliced
100ml/3½fl oz white vinegar
For the garnish:
lettuce leaves
225g/8oz white cheese such as Wensleydale, white Cheshire or Caerphilly, in 6 wedges
black olives

ALL THE BITS AND PIECES

Brazilians are economic with their produce: fresh beef may well be plentiful but in the national dish *Feijoada completa* you will find pigs' feet, ears and tails, smoked tongue and sausage and large amounts of black beans as well as the best beef available. It is traditionally served with rice and *farofa*, roasted cassava meal blended with palm oil.

- ***Serves 6*** 
- ***1235cals/5187kjs per serving***

1 Combine the finely chopped onion, lemon juice, cayenne pepper and salt and pepper to taste in a small bowl.

2 Put the potatoes in boiling, salted water, cover and simmer until they are tender, about 20 minutes. Drain them well and mash them.

3 Add 225ml/8fl oz of the olive oil to the reserved onion and lemon juice mixture. Pour this over the warm mashed potatoes, mixing thoroughly. Mound the potatoes in the centre of a large, round, heated serving platter and keep warm.

4 Put the sweet potatoes and cassava, if using, in boiling, salted water to cover and simmer for 20 minutes, or until tender, then drain and keep warm. It does not matter if the cassava slices break up.

5 Put the green bananas, unpeeled, but cut across into halves, if necessary, to fit into the pan, into a separate saucepan. Cover with boiling water and simmer about 15 minutes or until tender.

6 Drain, peel and cut the bananas into 12 slices. Keep them warm with the sweet potatoes and cassava.

7 Drop the corn into a large saucepan of boiling, salted water and boil for 5 minutes. Drain and cut each cob across into 3; keep warm with the vegetables.

8 Dredge the fish pieces in seasoned flour. Heat 50ml/2fl oz oil in a frying pan over medium heat and fry the fish pieces until they are golden brown on both sides, about 3-4 minutes. Drain on absorbent paper and keep warm.

9 Remove the stems and cut the chillies lengthways into strips about 3mm/⅛ in wide and drop the strips into a saucepan of boiling water with the sliced onions. Blanch for about 2 minutes, drain.

10 Return the chillies and onion slices to the pan, add the remaining olive oil, the vinegar, salt and pepper to taste and bring to a boil over low heat. Cook, covered, for 2 minutes.

11 To assemble, garnish the edge of the platter with lettuce leaves. Arrange the fish, corn, sweet potato, cassava, and bananas on the lettuce around the mashed potatoes and pour on the chilli and onion sauce. Garnish with the cheese and olives.

Cook's tips

Because this sauce is so fierce it might perhaps be wiser to serve it separately and let people test it out first.

Cassis and almond jelly

This delicate dessert, 'Blanc manger', made from ground almonds, is obviously derived from the original French *blanc-mange*, but Argentine cooks have added their own special touch

- ***Preparation: 30 minutes, plus chilling***
- ***Cooking: 20 minutes***

LATE ARRIVALS

After the Spanish and Portuguese came settlers from other parts of Europe: Italians settled in Argentina and Uruguay whilst many British and German immigrants live in Chile. With them came their own produce – strawberries, peaches and above all grapes. Chilean wine is now much appreciated and is exported all over the world.

500ml/18fl oz milk
75g/3oz almonds, finely ground
100g/4oz sugar
1tbls gelatine
½tsp almond essence
350ml/12fl oz thick cream
2tbls crème de cassis

- ***Serves 6*** 
- ***420cals/1765kjs per serving***

1 Combine the milk, almonds and sugar in a saucepan and simmer over low heat for 15 minutes. Strain through a sieve, pressing the almonds firmly with a wooden spoon to extract all the flavour.

2 Sprinkle the gelatine on to 50ml/2fl oz water in a small bowl and let it soften for 5 minutes. Stir the softened gelatine into the hot almond-cream mixture and continue stirring until dissolved. Stir in the almond essence. Allow to cool.

3 Whisk 125ml/4fl oz thick cream until stiff peaks form and fold it gently into the cooled almond mixture.

4 Rinse a 1L/1¾pt ring mould in cold water. Pour in the almond mixture and refrigerate until it has become firmly set.

5 Whisk the remaining cream until it is stiff and fold in the crème de cassis carefully in streaks.

6 Carefully unmould the almond jelly on to the centre of a flat serving dish and pile the cassis-flavoured cream into the centre.

Cook's tips

Fruit such as raspberries, strawberries or other berries may be added to the cream. Sliced peaches or apricots can be used.

Canadian Capers

True Canadian cooking is based on the old recipes of the earliest French settlers, combined with local produce

A baked ham loaf with a traditional mustard sauce (page 57)

AS FRANCE SOUGHT to extend its overseas empire, it sent settlers in the steps of Jacques Cartier, the explorer who landed in Canada in 1534 and claimed sovereignty for France. The early attempts at colonization were not successful and it was not until the early 17th century that French settlers founded what became the city of Quebec.

Early days

The first settlers were farmers and trappers, but they had to face a cruel, inhospitable land. They had never faced such hard winters, such bitter cold. Nothing grew in that harsh climate and the lack of vitamin C brought scurvy in its wake. Summer brought native fruit and vegetables and the settlers learnt to dry and store them – and as time passed, seeds were brought from France and welcome crops of peas, beans, oats and barley were grown. Maize, of course, was the native cereal, easy to grow and easy to store. Livestock was another problem. Cold winter weather demands a diet rich in animal protein and though the trappers were skilled men, it was not the same as hunting game in the fair woods and fields of France! They had to learn the native ways of trapping moose and caribou, hare, beaver and porcupine.

Fare for fasting

The new French-Canadians were devout Roman Catholics and always respected the many fast days prescribed by the church. Fish therefore was vital to their diet, as eggs and cheese were in very short supply. There was no shortage of sea and freshwater fish – salmon, haddock, sturgeon and eels were abundant. Once caught, they could be smoked and salted and stored away for those winter months. The fruits which grew so plentifully were also dried and stored so that the dreaded scurvy became a thing of the past.

Foreign invasions

By the end of the 18th century, thousands of immigrants from New England crossed the border to Canada, many of them with very English-style food and traditions, and as time went on more and more settlers left England and Scotland to seek a new life on the prairies of Canada. Since the end of the Second World War, Canada has welcomed thousands of immigrants from all parts of Europe so that Canadians can now eat as wide a variety of food as Americans – but the old cooking traditions are just as strong and the great dishes of the earliest days are favourites throughout the land.

Canadian fisherman drying fish on racks

SWEET SYRUP

Maple syrup, made from the sweet sap of the maple tree, is as typically Canadian as the Royal Canadian Mounted Police. It is traditionally served with pancakes and waffles, especially at breakfast time, and can be used to flavour ice-cream and to make delicious candy. Beware of imitations! Pure maple syrup is expensive and consists of 100% boiled-down sap. You can buy maple-flavoured syrup.

Breakfast yeast pancakes

- ***Preparation: 15–20 minutes***
- ***Cooking: 15 minutes***

½pkt easy-blend yeast
850ml/1½pt lukewarm milk
1tbls sugar
1egg
1½tsp ground cardamom
1½tsp salt
450g/1lb strong white flour
vegetable oil for greasing
butter, to serve
maple syrup and jam, to serve

- ***Serves 8***
- ***355cals/1490kjs per serving***

1 Sift the flour into a warmed bowl, add the yeast and mix. Heat the milk until just lukewarm.

2 Add the sugar, egg, cardamom and salt to the flour mixture. Pour on a little milk and beat well, add milk gradually until the batter is smooth.

3 Heat a griddle or heavy frying-pan over medium-high heat, using a second frying-pan as well if possible. Lightly grease the surface and pour in about 1½fl oz of the batter. When bubbles appear generally over the pancake, 30 seconds to 1 minute, flip it over with a spatula and cook the other side for 15–30 seconds.

4 Transfer the pancake to a warm plate, cover with a tea-towel and cook the other pancakes, putting them under the towel as they are cooked. Serve with butter and maple syrup or jam.

Canadian fruit pickle

- **Preparation: 30 minutes**
- **Cooking: 3 hours**

800g/1¾lb canned tomatoes
450g/1lb cooking pears
700g/1½lb cooking apples
4 onions, chopped
1 large red pepper, seeded and chopped
2tsp salt
2tbls pickling spice
200g/7oz sugar
300ml/½pt cider vinegar

- **Makes about 2L/3½pt** 
- **1440cals/6050kjs total**

1 Put the tomatoes and their juice into a large saucepan. Core and chop the pears and apples without peeling them and add them to the pan with the chopped onions and red pepper and salt.

2 Cook over low heat, uncovered, for 1½–2 hours or until reduced by half, stirring occasionally.

3 Tie the pickling spice in a piece of muslin. Add this with the sugar and vinegar to the pan. Cook uncovered for 1 hour, stirring often, until the mixture is as thick as chutney. Sterilize and warm pickling jars.

4 Remove the bag of pickling spice. Pour the mixture into the jars. Cover the surface of the pickle with circles of greaseproof paper. When cold, cover the jars tightly with vinegar-proof lids and store for up to 4 months.

Ham loaf with mustard sauce

- **Preparation: 25 minutes**
- **Cooking: 1 hour 10 minutes, plus 15 minutes resting**

700g/1½lb lean minced ham
350g/12oz lean minced pork
175g/6oz fresh breadcrumbs
125ml/4fl oz milk
2 eggs
about ½tsp salt
butter for greasing
125g/4oz canned tomatoes, drained and chopped
¾tsp sugar
celery leaf, to garnish

For the mustard sauce:
2½tbls flour
3 eggs
50g/2oz soft brown sugar
125ml/4½fl oz cider vinegar
175ml/6fl oz chicken stock
2tsp or more mustard powder

- **Serves 8** 
- **365cals/1535kjs per serving**

1 Heat the oven to 180C/350F/gas 4. Prepare the meats and mix the ingredients in a food processor if available. Combine the ham, pork, breadcrumbs, milk, eggs and salt in a bowl. Pour the mixture into a greased 1.7L/3pt loaf tin.

2 Combine the tomatoes and sugar in a small saucepan over low heat. Simmer for 5 minutes then pour over the meat loaf.

3 Bake the loaf for 30 minutes, then cover it with foil and continue baking until done, about 1 hour. Let the meat loaf rest in the tin for 10–15 minutes before turning out.

4 Meanwhile, prepare the sauce by whisking the flour into the eggs in the top of a double boiler. Stir in the sugar, vinegar, stock and dry mustard, adding more mustard if wished. Cook over simmering water until the sauce is thick and pour it into a warmed sauce-boat.

5 Serve the meat loaf in slices, garnished, pass the sauce separately.

Partridges with oyster stuffing

Canadians used to cook grouse and quail this way. It is still a good method of handling older game birds as the initial steaming keeps them moist and more tender. Mushrooms are sometimes used instead of oysters in the stuffing.

- ***Preparation: 1 hour***
- ***Cooking: 1 hour***

4 oven-ready partridges
8 slices rindless smoked streaky bacon
4tbls brandy
50g/2oz butter, softened
lemon wedges, to garnish
watercress, to garnish
For the stuffing:
225g/8oz canned oysters, drained
100g/4oz fresh white breadcrumbs
1tsp dried mixed herbs
large pinch of freshly grated nutmeg
1 egg, beaten
salt and freshly ground black pepper
50g/2oz butter, softened

- ***Serves 8***
- ***370cals/1555kjs per serving***

1 To make the stuffing, chop the drained oysters into small pieces and mix them with the breadcrumbs, herbs and nutmeg. Add the beaten egg and salt and pepper, and mix in the softened butter, using it to bind the stuffing.

2 Put an equal quantity of stuffing into each bird and truss them securely. Cut the bacon slices in half across and cover the breasts with them.

3 Put a trivet or rack in a saucepan or fish kettle large enough to hold the birds. Pour in just enough water to cover the rack. Put the trussed and stuffed partridges on the rack and bring the water to the boil.

4 Cover the pan, reduce the heat and steam the birds gently until just tender, 30–45 minutes depending on their age and size. Heat the oven to 200C/400F/ gas 6.

5 Transfer the birds, still on the trivet or rack, to a roasting tin. Lay the bacon rashers on the rack beside them. Warm the brandy slightly, set it alight and pour it over the birds. When the flames die down, smear the breasts with the butter.

6 Put the birds in the oven for 15 minutes or until well browned; baste them twice.

7 Split each bird in half along the breast and back bones. Place the halved birds, stuffing side down, on a warmed serving platter. Arrange the bacon and lemon wedges between them, garnish with watercress and serve hot.

Christmas Eve pie

Eaten by French Canadians on Christmas Eve, this pie is based on a very early dish, which used to be made of game.

- ***Preparation: 1 hour, plus 1 hour chilling pastry***
- ***Cooking: 1 hour 5 minutes***

25g/1oz unsalted butter
1medium-sized onion, finely chopped
1 stalk of celery, finely chopped
2 garlic cloves, finely chopped
450g/1lb lean minced pork
150ml/5fl oz chicken stock
½tsp salt
¼tsp dried thyme
¼tsp ground allspice
1 bay leaf
about 25g/1oz fresh white breadcrumbs
1tbls beaten egg or thin cream
For the pastry:
225g/8oz flour
½tsp salt
100g/4oz unsalted butter
50g/2oz lard

- ***Serves 4*** 
- ***730cals/3065kjs per serving***

1 First make the pastry. Mix the flour and salt, then cut in the butter and lard with a pastry blender or 2 knives until the mixture resembles oatmeal. Mix in enough cold water, about 3–4tbls, to make a dough which just holds together. Shape the dough into a ball and chill for 1 hour or longer.

2 Meanwhile, heat the butter in a heavy saucepan over medium-low heat and sauté the onion and celery in it for 4 minutes. Add the garlic and continue cooking for 2 minutes.

3 Add the minced pork and cook, stirring, until no trace of pink is left. Add

the stock, salt, thyme, allspice and the bay leaf. Cover, reduce the heat and simmer for 30 minutes.

4 Cool the mixture to tepid. Remove the bay leaf and mix in enough breadcrumbs to make the mixture moist but without free liquid.

5 Grease a 20cm/8in pie plate and heat the oven to 220C/425F/gas 7.

6 Roll out just over half the pastry into a 25cm/10in round on a lightly floured surface and line the pie plate. Turn the meat mixture into the lined plate and level the top.

7 Roll the remaining dough into a 21cm/8¼in round. Moisten the pastry rim and cover the pie. Press the pastry edges together and trim them.

8 Cut a cross in the centre of the pie to let steam escape. Make pastry strips from any trimmings and arrange them on the pie. Brush the egg or cream lightly over the pastry shell.

9 Bake the pie for 10 minutes, then reduce the heat to 190C/375F/gas 5 and bake for 25 minutes longer or until the crust is golden brown. Serve hot.

Cook's tips

This pie reheats well; store, covered, in the refrigerator. After Christmas make the pie with cooked minced turkey or game mixed with ham. Do not reheat the pie if using once-cooked meat.

WINTER FARE

Faced with only various dried pulses for their winter vegetables, the settlers took a leaf out of the Indians' book and made succotash: equal quantities of sweetcorn, broad beans and haricot beans are simmered in chicken stock with finely chopped onions, then seasoned well with salt, pepper and a little butter. French-Canadian pea soup is another dish with a long history. The addition of cornmeal gives it a unique flavour with extra body.

Planked fish

A seasoned oak or other hard-wood plank is a lovely means of cooking and serving fish and it adds good flavour. To season a new plank, brush it well with vegetable oil and heat it in an oven at 100C/225F/gas ¼ for 1–2 hours before using.

- ***Preparation: 15 minutes***
- ***Cooking: 40 minutes***

1.4kg/3lb whitefish, bass, salmon or other firm fleshed fish, scaled and cleaned
clarified butter or vegetable oil for greasing and brushing

- ***Serves 6*** 
- ***180cals/755kjs per serving***

1 Heat the oven to 180C/350F/gas 4. Wash the fish and dry it well with absorbent kitchen paper.

2 Put the fish on a seasoned, well greased plank that will hold the fish comfortably, or put it on a well greased baking tray. Brush the fish with clarified butter or vegetable oil.

3 When the oven is hot, bake the fish for about 40 minutes or until the flesh is tender but still moist, brushing it occasionally with the butter or oil. Serve immediately.

Cook's tips

Serve accompanied with lemon wedges and if liked Hollandaise sauce

Dried apple cake

- ***Preparation: 30 minutes, plus soaking time***
- ***Cooking: 2¼ hours, plus 10 minutes cooling***

175g/6oz dried apple rings
butter for greasing
175ml/6fl oz molasses
100g/4oz butter
2 eggs
225ml/8fl oz milk
175g/6oz soft brown sugar
1tsp bicarbonate of soda
225g/8oz flour, sifted
Glacé icing (see Cook's tips)

- ***Serves 8***
- ***485cals/2035kjs per serving***

1 Soak the apple rings in plenty of warm water overnight. Drain, reserving 175ml/6fl oz of the soaking liquid. Chop the apple rings roughly.

2 Put the chopped apple, 175ml/6fl oz of the soaking liquid and the molasses in a heavy saucepan. Mix well and simmer uncovered over low heat for ½ hour or until the liquid is reduced to a thin layer of syrupy glaze on the bottom of the pan.

3 Off the heat, beat in the butter with an electric beater until well blended. Heat the oven to 180C/350F/gas 4.

4 Beat the eggs into the milk and reserve. Mix the sugar, bicarbonate of soda and flour together in a bowl. Using one third of the dry mixture at a time, beat the dry ingredients into the molasses-butter mixture. Add the two mixtures together alternately with the egg-milk mixture until smooth.

5 Grease a 22cm/8½in round cake tin and stand a greased 450g/1lb jam jar in the middle. Half-fill the jar with water to make it heavier.

6 Carefully pour the mixture into the prepared tin without moving the jar. Level the top and bake for 1–1¼ hours or until the cake begins to shrink from the sides of the pan.

7 Let the cake cool in the tin for 10 minutes. Twist the jar and remove it. Leave the cake for another 10 minutes then carefully remove it from the tin. Finish cooling on a wire rack, if possible, for 24 hours. Ice before cutting.

Cook's tips

For glacé icing, sift 175g/6oz icing sugar; beat in about 2tbls warm water until it coats the spoon. Beat until smooth.

The Heart of China

Not only do we have the 'rice bowl of China' in the centre of the country but Szechuan and Hunan are renowned for the extra spiciness of the food

Ingredients for Crispy fish with pork and black beans (page 62)

TWO AREAS OF China are specially noted for the rich, hot spiciness of their cooking – Szechuan, which is almost in the centre of the country, and Hunan at the heart of 'the rice bowl of China'.

Hot stuff

It is the fiery red fagara chilli pepper which gives Szechuan cooking its spicy reputation – a fierce flavour which is actually supposed to make the taste buds more responsive to other flavours. Combined with vinegar and soy sauce, chilli produces a variety of multi-flavoured dishes to tease the palate. A popular Szechuan sauce is sold in jars – *tou pan jian*, a delicious mixture of soya paste, chopped ginger and garlic, onion and chilli. It adds terrific flavour to stir-fried food. Sometimes fish and poultry are plainly cooked, but they are served dressed with hot sesame sauce or with hot mustard sauce, excellent with shredded roast duck.

Smoking, Drying and Pickling

Because Szechuan is so far inland, fish and meat are often preserved by smoking and drying. Vegetables are pickled and are used on their own or mixed with shredded meat to give a typical Szechuan flavour. Chilli-hot mustard greens *(ja ts'ai li)* are a particular favourite. Added to fagara chilli, dried shrimps and mushrooms, these pickles are stirred with meat into noodles to make the popular *dan dan* noodle dish.

Slow spicy cooking

Hunan cooking also makes great use of chilli, vinegar and mustard, but its particular speciality is the *wei* method of cooking: cheaper cuts of meat are slowly casseroled to produce succulent tender joints and rich gravy. They are served sprinkled wth dried tangerine peel, chopped green and red peppers and a perhaps alarming amount of dried chilli and peppercorns.

Again, much of the meat and fish is smoked, salted or wind-dried, producing a distinctive Hunan flavour in local cooking. Smoked meats may be mixed subtly with stir-fried food and noodles, though they are most often served steamed.

All these methods produce food with a vigorous strong flavour and this is enhanced by the use of sesame paste and seeds, peanuts, sweet and sour sauces and home-made pickles. Tofu (bean curd) dishes are less bland and are enriched with chilli and garlic, soya paste and red bean curd cheese.

Transporting fruit in baskets

Crispy fish with pork and black beans

- ***Preparation: 30 minutes***
- ***Cooking: 25 minutes***

1kg/2¼lb whole firm white fish, such as grey mullet, sea bass, sea trout, bream, hake, brill, carp or turbot
2tbls cornflour
vegetable oil for deep-frying
5tbls oil
100g/4oz fat pork, cut into 5mm/¼in cubes
3tbls salted black beans
2 garlic cloves, roughly chopped
3 thin slices root ginger, roughly chopped
3tbls soy sauce
3tbls tomato purée
4tsp chilli sauce
4tsp sugar
4tbls good stock
5tbls Chinese rice wine or dry sherry
2 spring onions, roughly chopped

- ***Serves 4***
- ***630cals/2645kjs per serving***

1 Scale and clean the fish thoroughly if necessary. Slash each side of the fish 6 times with a sharp knife. Mix the cornflour with 4tbls cold water and reserve.

2 Heat 5cm/2in of vegetable oil in a deep frying pan large enough to hold the fish over medium heat. Fry the whole fish for 12-13 minutes until head, tail and protruding bones are crisp and golden on both sides. Lift the fish out carefully with one or two large fish slices and drain on absorbent paper.

SMOKED DELICACIES

Hunan smoked foods are first salted by being packed in salt, some sugar, peppercorns and wine; 6 or 7 days later they are taken out, brushed clean and hung to dry. Then they are smoked over sawdust and nutshells for 1-2 days. They turn a rich golden colour and have a flavour so intense that they must be rinsed under hot running water before being steamed or shredded for stir-frying.

Quick-fried duck with ginger and leeks

3 Heat the oil in a wok or a large, thin frying pan over high heat. When hot, add the pork and stir-fry it for 3 minutes. Add the black beans, garlic, ginger, soy sauce, tomato purée, chilli sauce, sugar, stock, wine and spring onions, stir-fry for 2 minutes or until the liquid has reduced by at least a quarter.

4 Put the fish in the pan, spooning over sauce. Arrange the fish on a heated serving dish.

5 Stir the cornflour mixture with the rest of the sauce for 1½ minutes or until the sauce has reduced by one-third. Pour the sauce over the fish and serve hot.

Quick-fried duck with ginger and leeks

Smoked foods are much used in Szechuan; substitute smoked chicken in this recipe if you wish

- ***Preparation: 20 minutes, plus 10 minutes soaking***
- ***Cooking: 6 minutes***

450g/1lb boned smoked or cold roast duck, cut into matchstick strips
4tsp salted black beans
4tbls vegetable oil
3 garlic cloves, roughly chopped
4 slices root ginger, cut into thin slivers
1 small red pepper, cut into matchstick strips
2 leeks, cut into matchstick strips
2tbls oil
4tbls good stock
2tbls soy sauce
2tbls distilled or white wine vinegar
2tsp sugar
2tsp chilli sauce

- ***Serves 5-6 as part of a larger meal***
- ***350cals/1470kjs per serving***

1 Soak the black beans in water to cover for 10 minutes and drain.

2 Heat the oil in a wok or a large thin frying pan over high heat. Add the beans, garlic, ginger, pepper and leeks; stir-fry for 2 minutes, then push round the sides of the wok.

3 Add the oil in the centre of the pan and add the duck and all the other ingredients and stir-fry vigorously for 2 minutes. Bring the pepper and leek mixture back to the centre of the pan and stir-fry them with the duck mixture for 1½ minutes. Shake the pan during cooking. Serve immediately with rice.

Steamed snails with ginger

- ***Preparation: 20 minutes***
- ***Cooking: 40 minutes***

2 spring onions
225g/8oz canned snails, drained
100g/4oz salt pork, ham or gammon, cut in strips
2 thin slices of fresh root ginger, cut in matchstick slivers
15g/½oz chilled butter, cubed
75ml/3fl oz good stock
¼tsp monosodium glutamate
2tsp oil
½tsp salt
boiled rice, to serve
mixed salt and pepper in equal parts, for dipping

- ***Serves 4-6 with other dishes***
- ***375cals/1575kjs per serving***

1 Cut the spring onions across in half and tie the green end of each one into a simple knot. Cut the remaining spring onion pieces into 5cm/2in sections.

2 Put the drained snails in a heatproof bowl. Spread the meat strips on top, add the ginger slices and spring onion knots, then dot the pieces of butter on top. Stand the bowl on a trivet in a steamer, bring the water below to the boil, cover and steam gently for 30 minutes.

3 Discard the spring onion knots and ginger. Transfer the rest to a bowl.

4 Put the stock, monosodium glutamate, the reserved spring onion pieces and oil in a small saucepan. Bring to the boil and season with salt. Pour over the snails.

5 Serve the dish, which is half soup, half stew, from the bowl with rice.

Hunan stuffed peppers

- ***Preparation: 20 minutes, plus 20 minutes soaking time***
- ***Cooking: 5 minutes frying, 25 minutes baking***

3tbls dried shrimps
10 dried Chinese black mushrooms
8 medium red peppers
250g/8oz lean minced pork
1½tsp salt
¼tsp pepper
pinch chilli pepper
2tbls soy sauce
2tbls cornflour
vegetable oil for deep frying
For the sauce:
2tbls oil
125ml/4fl oz stock
2tbls soy sauce
½tsp monosodium glutamate
4tsp oyster sauce
2tbls cornflour mixed with 3tbls water

- ***Serves 6-8 with other dishes***

- ***270cals/1135kjs per serving***

1 Soak the dried shrimps and mushrooms for 20 minutes in warm water.

2 Meanwhile, remove the stems, all the seeds and any pith carefully from the peppers through the hole at the top, making sure you keep the peppers whole.

3 Finely chop the pork or make filling in a food processor. Drain and chop the shrimps and mushrooms (discarding the hard stem tips) and add to the pork. Stir in the salt and soy sauce. Stuff the mixture in each pepper through the stem-hole, filling each one completely. Make the paste from the cornflour mixed with 2tbls water. Seal the stuffing at the top by spreading the paste over the hole. Do not allow to sink into pepper

4 Heat the oil in a deep-fat frier to 180C/350F; at this temperature a cube of stale bread will brown in 60 seconds. Lower the peppers, pasted-side down, into the oil one by one. Turn the heat to low and fry the peppers for 3-4½ minutes or until the skins are softened and are slightly blistered.

5 Remove the peppers from the oil and drain them. Arrange them, pasted-end upward, in a heatproof dish. Stand the dish on a trivet in a steamer and bring the water beneath to the boil. Cover and steam steadily for 15 minutes.

6 Alternatively brush the peppers over with oil and bake a hot oven 220C/450F/gas 8.

7 Meanwhile heat the oil in a frying pan over medium-low heat and add the rest of the sauce ingredients. Stir them, over high heat, for 1½-2 minutes until the sauce thickens. Pour the sauce over the stuffed peppers and serve immediately, from the heatproof dish.

SHOPPING SPREE

A visit to a Chinese supermarket can be baffling, but never fear to ask for what you want. You will easily find 'taste powder' (MSG or Ve-Tsin) on the shelves and there are all sorts of chilli sauces to try. Look for oyster sauce, delicious with stir-fried greens, or black bean sauce to blend with chicken or beef. And of course there are all the traditional Chinese pickles such as chilli bamboo shoots to liven up the noodles.

Hunan fish steaks

- ***Preparation: 20 minutes, plus 45 minutes soaking and marinating***
- ***Cooking: 20-25 minutes***

700-900g/1½-2lb firm white fish such as cod, sea bass or haddock, skinned
2tbls soy sauce
vegetable oil for deep frying
2½tbls chopped spring onions
2tsp sesame oil
boiled rice, to serve
For the sauce:
4 dried Chinese black mushrooms
2 onions, chopped
2 slices of bacon, rind removed
3 slices of root ginger, chopped
2 garlic cloves, chopped
1-2½tbls chopped ja ts'ai (radish pickle)
3-4 dried red chillies
150ml/5fl oz good stock
2½tbls soy sauce
50g/2oz sugar
2½tsp salt
4tbls rice wine or dry sherry

- ***Serves 4-5*** (fork & spoon) (££)
- ***605cals/2540kjs per serving***

1 Soak the mushrooms for the sauce in warm water for 30 minutes, drain and reserve them.

2 Meanwhile, clean the fish and cut into 2.5cm-4cm/1-1½in thick steaks. Rub them with the soy sauce and leave for 15 minutes.

3 Chop the onions, bacon, ginger, garlic, soaked mushrooms (discarding the stem tips), pickle and chillies for the sauce and reserve them separately.

4 Heat the oil in a wok or deep-fat frier to 180C/350F; at this temperature a cube of stale bread will brown in 60 seconds. Lower the fish steaks into the hot oil, a few at a time, to fry for 2½ minutes, reheat the oil before frying again. Drain on absorbent paper.

5 Heat 2½tbls vegetable oil in a wok or saucepan. Add the chopped sauce ingredients with the stock and soy sauce and stir-fry them together for 2 minutes. Add the rest of the sauce ingredients and stir until the sauce thickens.

6 Add the fish and baste the steaks with the sauce. Cover and cook over high heat for 4-5 minutes. The sauce should be reduced by half.

7 Transfer the fish to a large serving dish and pour the sauce over it, finally sprinkling the dish with the chopped spring onions and the sesame oil. Serve with a large bowl of boiled rice.

Chinese-style toffee apple fritters

This is an attractive dish to cook in front of your guests. You can fry the apple fritters in the kitchen and caramelize them at the table in a fondue pot. You must dip the fritters in iced water at the end or your guests will burn themselves. The Chinese call these 'silk thread' fruit because of the way in which the caramel pulls in threads from the fritters.

- ***Preparation: 30 minutes***
- ***Cooking: 40 minutes***

2 crisp dessert apples, preferably Cox's orange pippins
oil for deep frying
iced water for dipping
For the batter:
125g/4oz flour
1½tsp baking powder
150ml/5fl oz tepid water
For the caramel:
225g/8oz sugar

- ***Serves 4*** 
- ***475cals/1995kjs per serving***

1 Make the batter by sifting the flour and baking powder into a bowl and make a well in the centre. Gradually add the tepid water, stirring constantly to incorporate the flour. Mix well to form a smooth batter.

2 Make the caramel: place the sugar with 2-3tbls of water in a saucepan and bring slowly to the boil, shaking the pan. Cook to a hard ball stage (130C/260F) and remove from the heat. Heat the oil in a deep-fat fryer to 180C/350F.

3 Peel the apples, cut each apple into 8 sections and core. Coat a few sections of apple at a time with the batter and cook in the hot oil for 2-3 minutes. remove with a slotted spoon and drain on absorbent paper. Keep warm while cooking the remaining apple sections.

4 At the table (if you have cooked the actual fritters in the kitchen): when all the apples are cooked, reheat the sugar syrup over a table-top burner until it is very lightly caramelized. Place 2-3 apple fritters in the caramel, turning them until they are well coated and then dip immediately in iced water to cool them and set the caramel (tongs are vital for this). Place the caramelized fritters on an oiled serving dish. Serve immediately to your guests while you continue the process.

Cook's tips

The tricky part of this recipe is carmelising the fritters. However, as this is a favourite in Chinese restaurants, it is worth practising to master the technique. Dipping the fritters in the caramel must be done as quickly as possible as it will continue to cook, will become bitter and burn. If it is cooking too quickly, remove the saucepan from the heat, place on a heat resistant mat and continue to coat the fritters, returning the saucepan to the heat if necessary.

STEAMY PUDDINGS

There is a strong tradition of peasant cooking in Szechuan: one of the favourite traditional dishes is the Long-steamed pork pudding where layer after layer of tasty ingredients are spiced with chilli, sesame and soy sauces and chopped coriander, packed into covered containers and steamed for hours. These puddings are now part of Chinese *haute cuisine* and are more subtle in flavour and texture.

Cook Cantonese

Canton has far more to offer than just sweet and sour pork. Its cooking is diverse and distinctive, using the freshest ingredients and a subtle blend of flavours

Stir-fried squid in shrimp sauce (page 71)

CANTONESE COOKING IS a tradition that has grown in sophistication over more than 700 years. Geography and climate are favourable; the province where the cities of Canton and Hong Kong are located, Kwangtung, has the longest coastline of any Chinese province. Its climate is subtropical, with heavy rainfall half the year, so that two crops of rice can be grown annually and the people have enough to enjoy it twice a day. Beef is eaten more here than in other Chinese provinces. Besides pork, chicken, duck and goose, the Cantonese also enjoy crab, clams, conger eel, crayfish, prawns, squid, scallops and oysters, not to mention an enormous array of fish. The lychee is only one of many luscious native tropical fruits.

Rice with everything

There is a seemingly inexhaustible list of dishes to go with plain rice, which is eaten at every meal. While the preparation of dishes run the gamut of Chinese culinary techniques, stir-frying is the Cantonese speciality.

On the whole, Cantonese dishes are neither highly seasoned nor peppery hot. Rather, a harmonious blending of different flavours is the rule of thumb. Because the freshest, not to mention live, ingredients are plentiful, there is no need to use many spices. The addition of a few condiments is all that is needed to bring out the flavour of the main ingredients. Cantonese food served abroad, however, may taste insipid. When fresh ingredients, especially fish and seafood, are not readily available and frozen or canned substitutes are used, the results are understandably disappointing.

Chicken and sweetcorn soup

- ***Preparation: 30 minutes***
- ***Cooking: 1½ hours***

1tsp sesame oil
4cm/1½in piece of fresh root ginger, peeled, finely chopped
350g/12oz canned sweetcorn, drained
1tbls cornflour
2 eggs
salt and pepper
thinly cut strips of cucumber skin, to garnish
For the stock:
2 large chicken quarters
giblets from 1 chicken
1 large onion, quartered
2.5cm/1in piece of fresh root ginger, peeled
3-4 black peppercorns

- ***Serves 4-6*** 
- ***215cals/905kjs per serving***

1 First make the stock: put the chicken quarters, giblets, onion, ginger and peppercorns into a large saucepan.

2 Add 1.7L/3pt cold water and bring to the boil. Lower the heat, cover and simmer for 1-1¼ hours, removing the chicken portions when cooked, about 35 minutes. Remove the flesh from the bones and return the bones to the pan for the remaining cooking time. Shred the chicken flesh into small pieces and reserve.

3 Strain off the liquid and, if necessary, make up to 1.1L/2pt with water.

4 To make the soup, heat the sesame oil in a large saucepan, add the ginger and fry for 1-2 minutes, stirring constantly. Add the stock, chicken and sweetcorn. Bring to the boil, then lower the heat.

5 Blend the cornflour to a paste with 1tbls cold water, stir in a little of the hot soup, then add to the pan. Bring back to the boil, then simmer for 1-2 minutes, stirring.

6 Break the eggs into a small bowl, beat lightly, then pour into the soup in a thin, steady stream, stirring constantly. Season well, garnish and serve.

Cook's tips

When you buy a whole chicken and don't want to cook the giblets, freeze them so you'll have them on hand for soup or a stock. Or buy them fresh from your butcher.

Sesame prawn toasts

- ***Preparation: 10 minutes***
- ***Cooking: 20 minutes***

175g/6oz cooked peeled prawns
1tsp sesame oil
2 egg whites
2tsp cornflour
4 slices of white bread, crusts removed
2-3tbls sesame seeds
oil, for deep frying

- ***Makes 16 pieces***
- ***70cals/295kjs per slice***

1 Finely chop the prawns with a fork and stir in the sesame oil. Whisk the egg whites with the cornflour until frothy, then add to the prawns and stir well.

2 Spread the prawn mixture thickly onto the bread slices on one side only. Sprinkle generously with the sesame seeds.

3 Half-fill a deep-fat fryer or a heavy-based saucepan with oil and heat to 180C/350F or until a 1cm/½in cube of day-old white bread turns golden brown in 60 seconds when it is dropped into it.

4 Fry each slice individually for 2 minutes, prawn side up, until the bread is light golden. Remove with a fish slice, drain on absorbent paper and reserve for up to 30 minutes.

5 Just before serving, reheat the oil and return the slices to the pan, one at a time. Fry, turning once, until they are brown and crisp. Remove, drain on absorbent paper, and cut each piece into four strips. Serve at once.

DIM SUM

This Cantonese speciality has become a household word in many countries throughout the world. Westerners might consider these morsels to be hot hors d'oeuvres, but in China people often make a whole meal of them; there are restaurants specializing in *dim sum* that pride themselves on serving a different selection of these nibbles every day of the week.

The best-known dim sum are dumplings stuffed with pork, beef or seafood, then steamed or fried.

Beef and green peppers

- ***Preparation: 30 minutes, plus marinating***
- ***Cooking: 20 minutes***

350g/12oz rump steak, cut into thin 2.5cm x 5cm/1in x 2in strips
2tbls sesame oil
1½tbls soy sauce
1tbls dry sherry
1 large onion, finely chopped
2 green peppers, seeded and cut into thin strips
1 garlic clove, crushed
1tsp finely chopped fresh root ginger
2tsp cornflour
150ml/¼pt beef stock
50g/2oz mushrooms, sliced
salt and pepper
flat-leaved parsley and mushroom pinwheels, to garnish

- ***Serves 4*** 
- ***220cals/925kjs per serving***

1 Place the strips of steak in a large, shallow dish. Mix together 1tbls oil, the soy sauce and the sherry and pour over the strips of steak. Cover and leave to marinate for 2-3 hours, stirring once or twice.

2 Heat the remaining oil in a wok or large, heavy frying pan and, when very hot, add the onion. Stir-fry for 3 minutes or until softened, then add the peppers, garlic and ginger and continue to stir-fry for another 2 minutes. Remove the mixture from the wok with a slotted spoon and reserve.

3 Add the meat and marinade to the wok and stir-fry for 3-4 minutes. Blend the cornflour with a little of the stock. Add to the beef, stirring it in, then add the remaining stock and stir well. Mix in the reserved pepper mixture plus the mushrooms, and season to taste. Simmer for 5-6 minutes or until the beef is tender. Transfer to a warmed serving dish and garnish with flat-leaved parsley and mushroom pinwheels.

Cook's tips

To make mushroom pinwheels, wipe some firm button mushrooms and trace a simple pattern with the point of a knife, peeling away the skin from the patterned area.

Stir-fried vegetables

- ***Preparation: 30 minutes***
- ***Cooking: 15 minutes***

3tbls sesame oil
1 garlic clove
1 onion, sliced
2.5cm/1in piece of fresh root ginger, peeled and finely chopped
2 celery stalks, sliced
1 large carrot, thinly sliced
1 small red pepper, cut into squares
1 small yellow pepper, cut into squares
1 small mooli, cut into matchstick strips
100g/4oz broccoli, separated into florets, thick stalks sliced
1 leek, cut into rings
50g/2oz button mushrooms, sliced
50g/2oz mange tout
4 asparagus spears, cut into 5cm/2in lengths
¼ head of Chinese leaves, sliced
2tbls light soy sauce
salt

◀ • *Serves 6*

• ***115cals/485kjs per serving***

1 Heat the oil with the garlic in a wok or large, heavy pan. When the oil is very hot, remove the garlic with a slotted spoon, add the onion and ginger and stir-fry for 2 minutes or until softened.

2 Add the celery, carrot, peppers, mooli, broccoli stalks and leek and stir-fry for 3-4 minutes. Add the mushrooms, broccoli florets, mange tout and asparagus spears and stir-fry for another 2-3 minutes.

3 Add the Chinese leaves and soy sauce and stir-fry for 1-2 minutes. Season with salt and serve.

Variations

You don't have to use all the vegetables listed in this recipe – mix and match with what is available, but always try to choose vegetables which complement each other in flavour, colour and texture.

Street food of all kinds is very popular in China. Here, hungry passers-by are tempted by noodles

WOK SENSE

Before using a wok for the first time, heat it over a high heat, then brush it lightly with oil. Wipe with absorbent paper before repeating twice. Rinse well and dry thoroughly.

Your wok may rust if not in constant use. If it does, scour the rust off, rinse and brush again with oil to re-season it.

Steamed scallops with black beans

• ***Preparation: 10 minutes***

• ***Cooking: 10 minutes***

8 scallops, with half shells
1tsp cornflour
2tbls oyster sauce
3tbls dry sherry
2tbls oil
2tbls canned black beans, drained
1cm/½in piece of fresh root ginger, peeled and very finely chopped
2 spring onions, finely chopped
8 cucumber fans, to garnish

• ***Serves 4***

• ***165cals/695kjs per serving***

1 Place the scallops and their corals on the half shells. Place in a steamer, cover and steam over boiling water for 6-8 minutes.

2 Meanwhile, make the sauce: in a small bowl or cup, mix together the cornflour, oyster sauce and sherry to form a smooth paste. Reserve.

3 Heat the oil in a wok or frying pan, add the black beans, ginger and spring onions and stir-fry for 1 minute. Stir in the cornflour paste and cook, stirring, for 1 minute.

4 Pour the sauce over the scallops and serve immediately, garnished with cucumber fans.

Making a cucumber fan

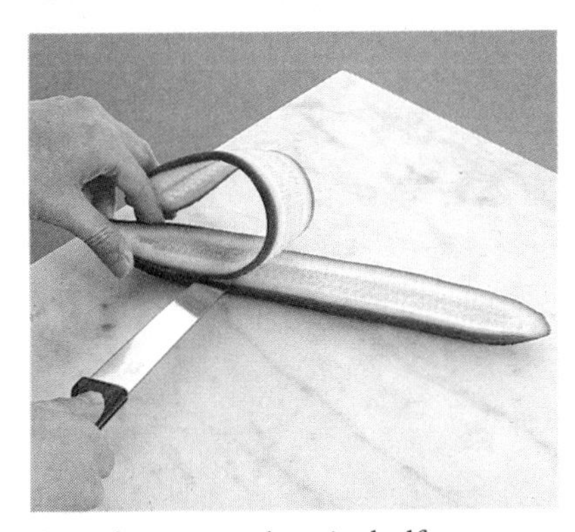

Cut the cucumber in half lengthways, then cut off a 5mm/¼in slice. Trim off the end.

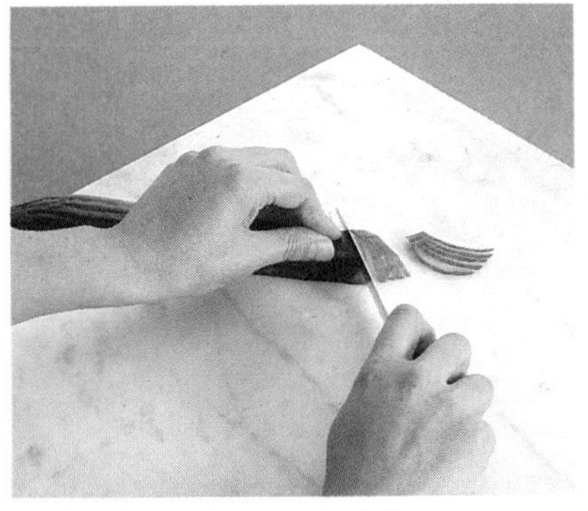

Make four close parallel cuts across but not right through, then cut off this section.

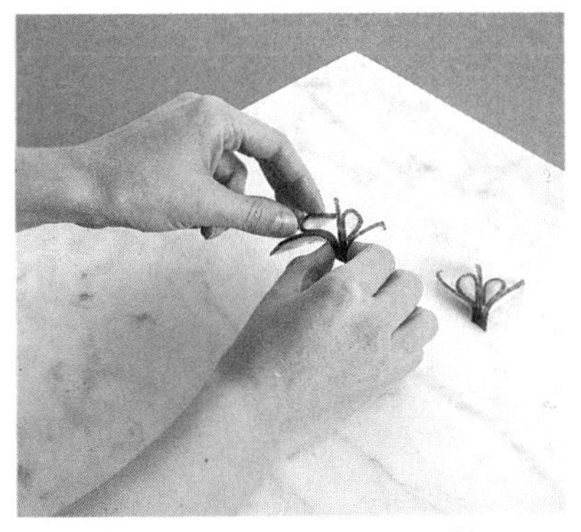

Repeat seven more times. For each fan, bend two of the slices towards the centre.

CHINESE FOODIES

Food is one of the great joys in life for all Chinese people, and particularly so for the Cantonese. They have a special liking for an exotic 'gamey' taste and enjoy flesh such as stewed cat, armadillo and snake as well as dog. These curiosities are not only considered gourmet delights but are reputed for their health-giving and aphrodisiac effects.

Stir-fried squid in shrimp sauce

- ***Preparation: 35 minutes***
- ***Cooking: 10 minutes***

450g/1lb small squid, cleaned
boiling water
600ml/1pt oil
1½tbls shrimp paste
3-4 garlic cloves, finely chopped
1cm/½in piece of fresh root ginger, peeled and finely chopped
6 spring onions, sliced thinly on the diagonal
1tbls medium-dry sherry
For the sauce:
2tsp dark soy sauce
1tsp tapioca or potato flour
4tbls chicken stock or water

- ***Serves 4***
- ***290cals/1220kjs per serving***

1 Skin the squid and lay on a board. Cut the body into 5mm/¼in rings and the tentacles into 4cm/1½in lengths, taking care to remove the bone and to wash away the black ink.

2 Immerse the cut squid and tentacles in plenty of boiling water for 10-20 seconds. As soon as they curl up, drain in a colander and then rinse under cold running water. Drain thoroughly and dry well with absorbent paper.

3 Heat the oil in a deep, heavy-based pan over high heat until it reaches 150C/300F; at this temperature a 1cm/½in cube of day-old white bread will brown in 1½ minutes. Add the squid pieces and cook for about 30 seconds or until turning golden. Remove the squid with a slotted spoon and reserve. Pour off all but 2tbls of the oil, keeping the rest for another recipe.

4 Dilute the shrimp paste with 1tbls water, stirring to blend; reserve. Mix the sauce ingredients together and keep tc one side.

5 Heat the reserved oil in a wok until it smokes. Add the garlic and, when it sizzles, add the ginger. Stir a few times and add the spring onions. Cook for 1-2 minutes or until the garlic turns golden brown. Stir several more times, then pour in the shrimp paste. Stir and cook for several seconds.

6 Return the squid to the wok. Toss the ingredients carefully from the bottom of the wok for 20-30 seconds or until hot. Stir in the sherry. When the sizzling subsides, add the sauce ingredients and continue to stir as it thickens.

7 Turn the contents of the wok out onto a warmed serving plate and serve.

Cook's tips

If you don't like the idea of preparing the squid yourself (though really it is very easy), you can sometimes buy it ready prepared in fishmongers.

You can buy shrimp paste in jars or dried in blocks from Oriental supermarkets, where it may be labelled blachen.

LUSCIOUS LYCHEES

Lychees were so loved by Yang Kuei-fei, the Imperial Concubine of Emperor Hsüan Tsung, who was the last great ruler of the T'ang dynasty, that they became a romantic symbol for the Chinese.

Sweet-scented lychees

- **Preparation: 15 minutes**
- **Cooking: 10 minutes, plus 2 hours chilling**

75g/3oz sugar
1tbls ginger syrup
grated zest and juice of 1 lemon
75ml/3fl oz dry white wine
450g/1lb lychees, peeled and stoned
2 pieces of stem ginger, chopped
1tbls finely grated lime zest

- **Serves 4** 
- **170cals/715kjs per serving**

1 Mix the sugar with 75ml/3fl oz water in a saucepan and place over low heat for 3-5 minutes or until the sugar dissolves, stirring constantly.

2 Add the ginger syrup, half the lemon zest, the lemon juice and wine, and bring to the boil. Add the peeled and stoned lychees, then reduce the heat to low and simmer, uncovered, for 5 minutes.

3 Remove the pan from the heat and transfer the lychees and syrup to a serving bowl. Set aside to cool before chilling for 2 hours.

4 Stir in the chopped ginger just before serving and sprinkle with the lime zest and remaining lemon zest.

CHOP SUEY

In 1842 the city of Canton was opened to the West; after this many Cantonese emigrated to Australia and California. Some of them set up eating houses and restaurants serving stir-fried bits of meat and vegetables, seasoned with soy sauce and called 'chop suey'. Many Westerners still regard this dish as native Chinese, although it is really a hybrid dish, created by the Cantonese in 19th century America.

Made in Japan

The Japanese are masters in the art of food: they believe it should be exquisite both to the tastebuds and to the eye

Sushi *(page 78)*

JAPANESE FOOD IS elegantly simple. It is characterized by a natural taste and a strong feeling for using foods in their season. The cooking methods used enhance the natural qualities of the ingredients and the presentation in bowls or on square or oblong dishes is designed to show off the food.

There is also humour in the naming of dishes: a rice dish with chicken and eggs is called 'parents and children' while another rice dish with beef and eggs translates as 'strangers and children' – a charming, if mild, culinary joke.

Until fairly recently, food was cooked over charcoal, and techniques, different from our usual chopping and slicing, were developed to cut ingredients thinly. The food then cooks quickly as a greater surface is exposed to the heat. Since all food is eaten with wooden chopsticks, it is cut into bite-sized pieces which are pretty to look at as well as easy to eat.

Foreign influences

The cuisine developed during long periods of isolation from the rest of the world, giving it a highly individual stamp. Certain foreign influences, though, have left their mark. From China came chopsticks and soy sauce and, in the 13th century, Zen Buddhism. This religion insisted on strict vegetarianism and this more or less prevailed until the 19th and 20th centuries when influence from the West in general, and France in particular, brought

Tea drinking

Tea and sake are the most common drinks to have with a Japanese meal. Green tea, or *nihon-cha,* is the tea drunk before, during and after any meal. It is green because it is made from young leaves which have not been processed. The famous Tea Ceremony of Japan is altogether different: for this there is special tea, a whole ritual of service and a category of food called *kaiseki* which accompanies the ceremony.

Geisha taking part in the ritualistic Tea Ceremony

popularity to meat and fish dishes. Portugal introduced battered, fried foods, adopted by the Japanese as *tempura,* and various vegetables such as the potato, sweet potato and pumpkin. Essentially, though, the Japanese kitchen has retained what it has had from the beginning – natural foods gracefully presented.

Eating Japanese

Many one-pot dishes, such as *sukiyaki,* are cooked at the table and served with rice, and may provide a whole meal. Sukiyaki is probably the best-known Japanese dish.

For a main meal there is always rice, a soup, a vegetable salad or a pickle dish, a fish or shellfish dish, and a meat or poultry dish, all served in small portions. The food is presented all at once and not eaten in any particular order. There are a few desserts but they are not intrinsic to the Japanese kitchen.

Sake, Japanese rice wine, is served warm in tiny sake cups, and may accompany the appetizers or be served throughout a festive meal. Wine is becoming increasingly popular, and dry white wine is probably the best choice. There is also excellent beer in Japan, and lovely, delicate green unfermented tea served plain in tiny bowls.

Clear soup with bean curd

- ***Preparation: making dashi, then 10 minutes***
- ***Cooking: 10 minutes***

1.1L/2pt dashi (see right)
salt
1tsp Japanese soy sauce
100g/4oz bean curd
6 mange tout, trimmed
1 lemon, thinly sliced

- ***Serves 6*** (!) (££)
- ***15cals/65kjs per serving***

1 In a large saucepan, slowly bring the dashi to the boil. Lower the heat, add a pinch of salt and the soy sauce and simmer for 3-5 minutes.

2 Cut the bean curd into 12 even-sized squares. Remove the pan from the heat, add the mange tout and bean curd squares and leave for 5 minutes, to allow them to heat through.

3 Place a slice of lemon in each soup bowl. Remove the bean curd and mange tout from the stock, using a slotted spoon, and put two pieces of bean curd and one mange tout in each of the bowls. Reheat the stock.

4 Fill each bowl with the soup stock and serve immediately.

Cook's tips

Bean curd has a delicate, soft texture and easily disintegrates, so try to avoid stirring the soup or bringing it to the boil once the bean curd has been added.

Its delicate flavour readily absorbs other, stronger flavours.

DASHI

Dashi is a fish and seaweed stock. The seaweed is called kombu (kelp) and the fish is finely flaked dried bonito, a fish similar to mackerel. Packet dashi is available from Oriental food shops and, like most packaged Japanese foodstuffs, is very high quality.

Vegetable tempura

The batter used in this recipe is a far cry from the stodgy Western sort. It is made with iced water, eggs and flour, and gives a light, almost transparent coating to the food

- ***Preparation: 50 minutes***
- ***Cooking: 25 minutes***

8 baby sweetcorn
1 small aubergine, halved lengthways
1 green pepper, halved and seeded
12 large button mushrooms, stalks removed
450ml/16fl oz oil
1tbls sesame oil
100g/4oz French beans
1 large carrot, cut into 7.5cm/3in sticks
flour, for dusting
watercress sprigs and lemon twists, to garnish
Tempura dipping sauce or soy sauce, to serve
For the batter:
300ml/½pt iced water
1 egg
100g/4oz flour
pinch of bicarbonate of soda

- ***Serves 6*** 
- ***350cals/1470kjs per serving***

1 First prepare the vegetables: cook the baby sweetcorn in boiling water for 2-3 minutes, then drain and pat dry on absorbent paper and set aside.

2 Scoop out the aubergine halves, leaving 1cm/½in thick shells. Cut each half into eight slices. Make parallel cuts in each slice, to form fans.

3 Cut the green pepper into diamond shapes, about 2cm/¾in wide and 4cm/1½in long. Cut cross-shaped wedges out of each mushroom cap.

4 Next make the batter: pour the iced water into a large bowl, add the egg and beat with a whisk until light and frothy. Lightly mix in the flour and bicarbonate of soda.

5 To cook the tempura, heat the oils in a tempura pan or deep-fat fryer to 170C/325F. (At this temperature a 1cm/½in square of day-old white bread will turn golden in 75 seconds.) Dust the well-dried vegetables in a little flour, shaking off any excess.

6 Dip the vegetables in the batter and deep-fry in batches for 2-3 minutes, turning once. Cook the baby sweetcorn first, then the carrots and beans in clusters.

7 Cook the peppers, aubergine fans and finally the mushrooms. Drain on absorbent paper and keep warm in a low oven.

8 Arrange the cooked vegetables decoratively on a serving tray. Garnish with watercress and lemon twists and serve immediately, with Tempura dipping sauce or soy sauce.

TEMPURA DIPPING SAUCE

Heat 3tbls mirin in a small pan. When it is about to boil, remove from the heat and set alight. Swirl gently, then allow the flame to die. Return to the heat and add 3tbls soy sauce and 300ml/½pt dashi. Bring to the boil, then cool. Put a mound of grated daikon in the centre of each of six small bowls and top with a little freshly grated ginger root. Spoon the cooled sauce around and serve at once.

Cook's tips

Do not overbeat the batter: there should be a fine spray of flour around the side of the bowl and on top of the batter; and the batter should look lumpy and not properly mixed.

Serving ideas

In a Japanese restaurant you are likely to be served tempura beautifully arranged on folded paper on a wooden tray.

Sukiyaki

- ***Preparation: 50 minutes, plus freezing***
- ***Cooking: 15 minutes***

450g/1lb sirloin or fillet steak
10 spring onions
250g/9oz drained weight shirataki, or 175g/6oz dried vermicelli
100g/4oz bean curd, cut into 12 even-sized cubes
100g/4oz shungiku (Japanese chrysanthemum leaves) or 225g/8oz spinach
1 onion, cut into thin wedges
1 bunch Chinese leaves, shredded, or watercress
225g/8oz canned bamboo shoots, drained and sliced
8 button mushrooms, sliced
125ml/4fl oz soy sauce
100ml/3½fl oz mirin (see Cook's tips)
100ml/3½fl oz dashi (see page 74)
pinch of monosodium glutamate (optional)
1tbls caster sugar
25g/1oz beef suet or 1tbls oil
4 eggs (optional)
rice (optional)

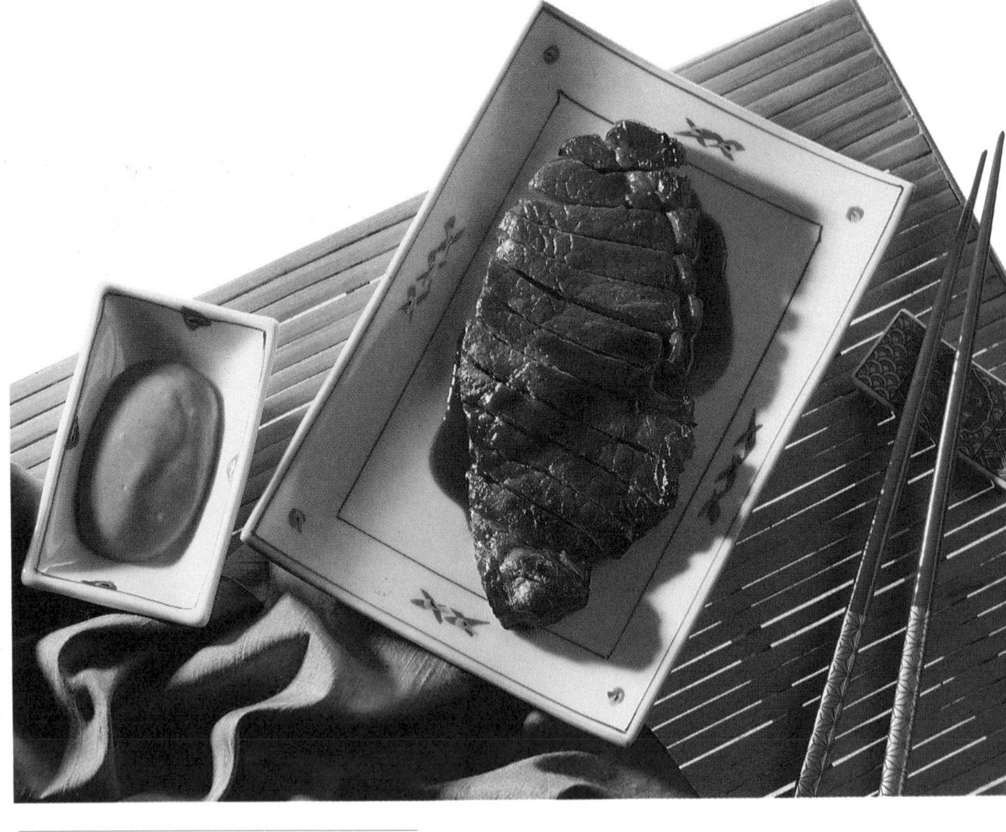

- ***Serves 4*** 
- ***490cals/2060kjs per serving***

Teriyaki steak

1 Freeze the steak for 30 minutes or just long enough to stiffen for easy slicing. Cut into very thin slices.

2 Meanwhile, trim the spring onions, then slice into 5cm/2in diagonal slices with a sharp knife.

3 Cook the shirataki in boiling water for 1 minute, drain and cut in half, or boil the vermicelli for 2 minutes and drain. Arrange with the steak, bean curd and vegetables on a platter.

4 In a small saucepan combine the soy sauce, mirin, dashi, monosodium glutamate, if using, and sugar. Bring to the boil, then pour into a small jug.

5 When ready to cook at the table, set a heavy based frying pan, about 25cm-30cm/10in-12in in diameter, over a table burner and let it heat for several minutes. Rub the pan with the suet or swirl round the oil.

6 To cook, add the ingredients to the pan a little at a time: add a little beef and cook for 1-2 minutes, turning once. Add the spring onions and onion, then pour half of the soy sauce mixture over the pan ingredients. Add half of the noodles, bean curd, green vegetables, bamboo shoots and mushrooms and cook, stirring with chopsticks, for 3-4 minutes. Serve this as soon as it is cooked, with rice if wished, then continue to add ingredients and cook as above.

7 If wished, break the eggs into four small bowls and stir. Everyone lifts out pieces of the ingredients and dips them into the egg before eating.

Cook's tips

Mirin is a sweet, low-alcohol cooking wine; sherry can be substituted.

Teriyaki steak

Here, teriyaki sauce is used to marinate and baste the meat, to make it both tender and tasty

- ***Preparation: 10 minutes, plus marinating***
- ***Cooking: 10-20 minutes***

4 × 175g/6oz rump steaks, trimmed of all fat
mustard, to serve (optional)
For the marinade:
2.5cm/1in fresh root ginger, peeled and finely chopped
2 garlic cloves, crushed
4 spring onions, finely chopped
25g/1oz soft brown sugar
125ml/4fl oz soy sauce
125ml/4fl oz sake or dry sherry

- ***Serves 4*** 
- ***365cals/1535kjs per serving***

1 Combine the marinade ingredients in a large shallow dish and place the steaks in the mixture. Set aside to marinate at room temperature for 2-3 hours.

2 Heat the grill to high. Transfer the steak to the grill pan, reserving the marinade. Brush the steak with the marinade.

3 Grill for 4-5 minutes. Remove the pan from the heat, turn the steak and brush with a little more of the marinade. Return to the heat and grill for a further 3-4 minutes for rare steaks. Increase the cooking time for medium and well done.

4 Meanwhile, bring the remaining marinade to the boil in a small pan and simmer gently while the steaks cook.

5 Slice the steaks, then arrange on heated serving dishes. Pour over a little of the marinade and serve at once, with mustard, if wished.

Japanese salad

- ***Preparation: 30 minutes***
- ***Cooking: 15 minutes, plus cooling***

50g/2oz soba noodles (see Cook's tips)
6 small asparagus spears
1 iceberg lettuce, shredded
1 carrot, cut into matchstick strips
6 radishes, thinly sliced
6 cucumber fans and flat-leaved parsley sprigs, to garnish
For the dressing:
2tbls rice or cider vinegar
2tsp caster sugar
2tsp cornflour
7tbls dashi (see page 74)
1tbls soy sauce
1 spring onion, finely chopped

- ***Serves 6*** 
- ***70cals/295kjs per serving***

1 First make the dressing: mix the vinegar with the sugar and cornflour in a small saucepan. Add the dashi, soy sauce and spring onion and slowly bring to the boil. Simmer for a few minutes, until slightly thickened. Remove from the heat and allow to cool.

2 Cook the soba noodles in boiling water for 3 minutes or until tender but still firm. Drain well and rinse under cold running water. Set aside.

3 Cook the asparagus in boiling water for about 5 minutes or until tender. Drain and refresh under cold running water. Using a sharp knife, cut each spear in half lengthways and leave to cool.

4 On six individual salad plates, arrange the lettuce, soba noodles, asparagus, carrot and radish slices. Pour the cooled salad dressing over. Garnish and serve.

Cook's tips

Soba noodles are one of the most popular varieties of noodle used in Japan. They only take a few minutes to cook, so be careful not to overcook them. They are often used in salads or as a substitute for rice.

METHODICAL COOKING

The Japanese meal structure is different from the Western pattern, the food being classified not by its place in the meal but by the cooking method. For example, *yakimono* are grilled foods, *gohan* are rice dishes and *mushimono* are steamed foods, while *sashimi*, sliced raw fish, is not cooked at all.

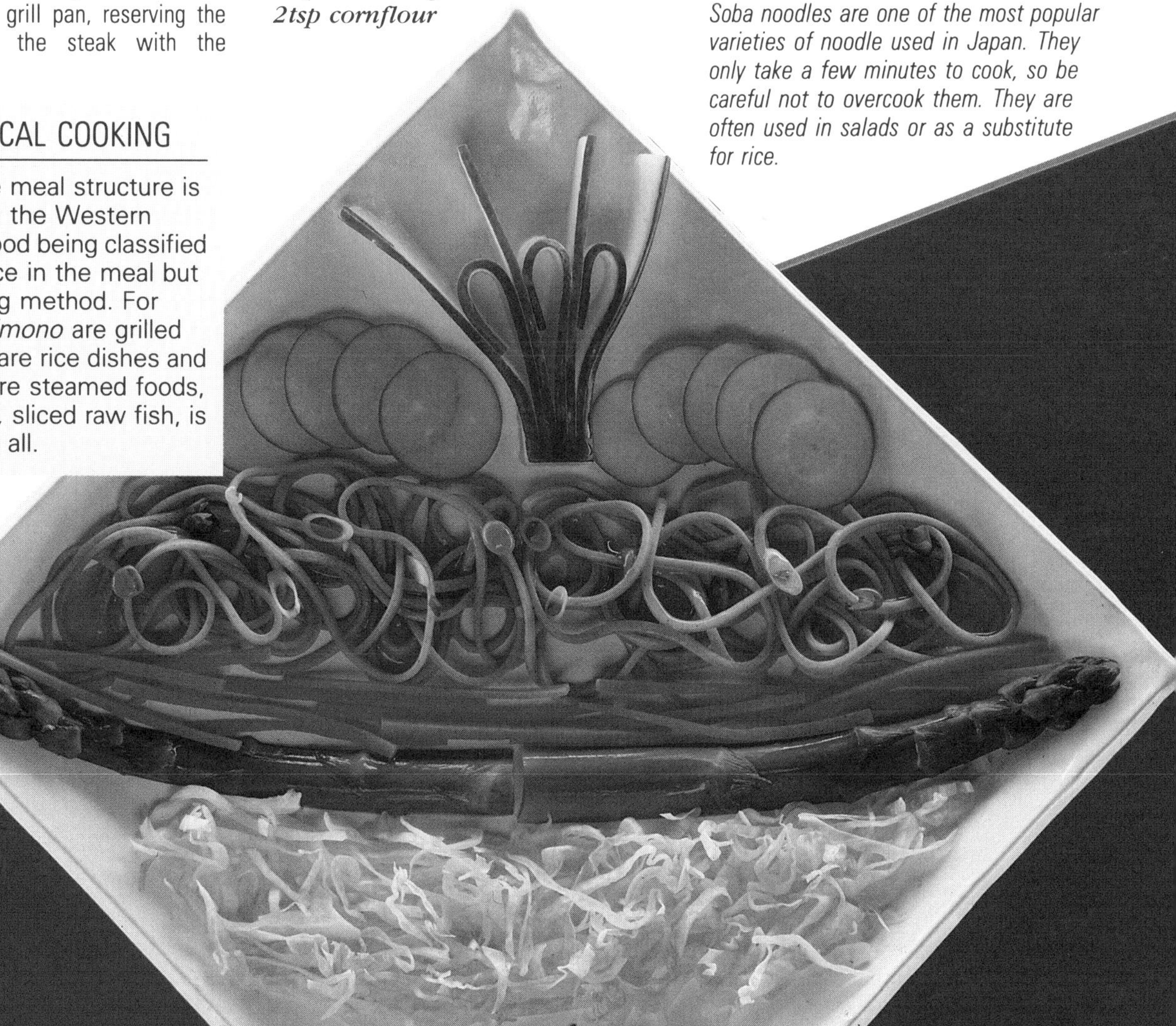

Sushi

Sushi is lightly vinegared rice, garnished with raw or cooked seafood. Each sushi should be a delight to the eye

- ***Preparation: 1 hour, plus draining***
- ***Cooking: 30 minutes, plus cooling***

For the rice:
400g/14oz short-grain rice
7.5cm/3in square kelp (kombu), cut into a 1cm/½in fringe
50ml/2fl oz rice vinegar, plus extra
1tbls sugar
salt
½tsp monosodium glutamate (optional)
For the topping:
1tsp rice vinegar
4 large fresh raw prawns
4 fresh scallops
100g/4oz fresh sea bass, sea bream or tuna
1tbls green horseradish powder, (wasabi), mixed with cold water to a thick paste
100g/4oz salmon roes
4tbls red caviar, real or mock
To garnish and serve:
sliced tomato, cut radishes, cucumber strips, onion tassels and capers (optional)
Japanese soy sauce

- ***Serves 4-6*** 
- ***510cals/2140kjs per serving***

1 Thoroughly wash the rice in several changes of water until it runs clear, then drain in a sieve for 1 hour.

2 Put the rice into a heavy saucepan with a tightly fitting lid. Bury the kelp in the rice. Add 500ml/18fl oz water, cover and bring to the boil over a high heat. Remove the kelp just before the water boils and re-cover.

3 Reduce the heat to medium and cook for 6-7 minutes. Then reduce the heat to very low and cook for 15 minutes. Raise the heat to high for 10 seconds, then remove the pan from heat. Let the rice stand, covered, for 10-15 minutes.

4 Combine the vinegar, sugar, 2tsp salt and the monosodium glutamate, if using, in a small saucepan over moderate heat and heat through until the sugar and salt have dissolved. Turn the rice out onto a large shallow dish, preferably wooden. Little by little, pour the vinegar mixture over the rice, mixing it with a wooden spatula or a fork, and fanning it vigorously at the same time. (It helps to have an assistant for this.)

5 When the rice is cool, wet your hands with water to which a little rice vinegar has been added, and form the rice into 20-24 oblong patties, about 2.5cm x 5cm/1in x 2in.

6 To prepare the topping, bring a small pan of boiling salted water to the boil. Add the vinegar and simmer the prawns for 1 minute. Drain immediately. When cool enough to handle, peel and cut the underside from end to end three-quarters of the way through. Then turn them over and flatten them.

7 Slice the scallop corals horizontally. Slice the discs into three horizontally. Slice the raw fish diagonally into slices 5mm/¼in thick.

8 Spread a dab of wasabi paste down the centre of each piece of fish and lay them, wasabi-side down, on top of some patties. Spread the salmon roes directly on some patties and put a dab of paste on top of the roes. Arrange the prawns on more patties. Arrange the kelp fringe on the side of some patties. Garnish, if wished.

9 Fill a small bowl with soy sauce and serve with the patties. To eat, use chopsticks or fingers to dip the patties into the soy sauce before eating.

Cook's tips

Make sure the fish and shellfish you use are very fresh, as they are to be eaten raw.

BUYING JAPANESE

Ingredients such as kombu, wasabi, mirin, rice vinegar, daikon (a large, mild radish also known as mooli), and shirataki (thin, translucent noodles) are available from Japanese or other Oriental food shops.

Orange jewel caskets

Japanese cuisine has very few desserts but attractively presented fresh fruit is a popular choice

- ***Preparation: 50 minutes, plus chilling***

6 oranges
225g/8oz black grapes, halved and seeded

- ***Serves 6***
- ***75cals/315kjs per serving***

1 Using a well-sharpened knife, cut the top off each of the oranges, creating a serrated edge pattern at the same time (see Cook's tips).

2 Cut the flesh from inside the orange by cutting all the way around, as close to the pith as possible. Carefully remove the ball of flesh. Remove the membranes, cut the flesh into bite-sized pieces and reserve.

3 Fill the orange shells with the reserved orange pieces and grape halves. Chill for at least 30 minutes before serving.

Cook's tips

For the attractive zig-zag edge, use a long thin knife with a sharp serrated edge and make an insertion, at a 45° angle to the base, about one-third of the way down the orange. Cut through the flesh as well, as far as it will go. Now make alternate cuts at 45° angles all the way around the orange until the top lifts freely away.

Eastern Crossroads

The Malaysian peninsula and the many islands of Indonesia mingle strong and subtle flavours to satisfy the most exacting palate

A typical Far Eastern spread: Prawn and vermicelli soup (page 80) and Crab with chilli (page 81)

WHEREVER THE CHINESE have travelled, they have taken along their cooking traditions – and nowhere illustrates this quite so well as the Malay peninsula and Singapore. Here, they settled among the native Malays and the traders from India and Ceylon to produce a hot and spicy version of Chinese food which is quite distinctive in the Far East. Hot chillies are a typical feature of local cooking and when added to a Chinese crab dish produce something deliciously spicy and hot (Crab with chilli, page 81). Delicate Chinese soups are transformed when Malaysian spice combinations and coconut milk are part of the recipe (Prawn and vermicelli soup, page 80).

Typically Malay

Perhaps the most typical Malay dish is satay (page 81), now one of the popular dishes on the Malaysian menu in Britian. In Singapore, it is sold from mobile street stalls together with bowls of fried rice noodles. The humid tropical climate produces lush fruit and vegetables, some of them now available in Far Eastern foodshops in Britain – rambutan, mangosteen, yard-long beans, fresh water chestnuts, bamboo shoots and breadfruits. Lemon grass is a favourite herb, together with fresh ginger and spicy-sour tamarind. And, of course, coconut milk, both thick and thin, to lend body and subtlety of flavour to meat and curry dishes.

Spice Islands

Indonesia is a vast archipelago of islands: over 3000 of them are inhabited and among them are the famous Spice Islands where spices have been bought and sold for centuries. These are the flavours which so tantalised Europeans when they ate native dishes. Add to these the powerful flavours of garlic, chilli and soya bean *kecap* (known to us as ketchup) and you have an idea of Indonesian cooking. Rice forms the basis of most meals and it is customary to put all the dishes on the table at the same time so that diners can help themselves and pile savoury meat, fish and vegetables on to a bed of rice. There will be side dishes of hot sambals, crunchy *ikan bilis* (dried salted anchovies), shrimp paste and chilli relishes – and at the end of the meal, fresh fruit and the offer of Java tea with masses of sugar and no milk.

Night-time market in Singapore's crowded Chinatown

Spicy meat soup

This recipe is well known in many parts of Indonesia. It may have originated on the island of Madura, off north-west Java. Soto is a general term for soup made with plenty of meat – so much meat that the soup seems more like a stew.

- ***Preparation: 20 minutes***
- ***Cooking: 1½ hours***

450g/1lb beef brisket
100g/4oz raw or cooked prawns, peeled
6 shallots
3 garlic cloves
2tbls vegetable oil
½tsp ground ginger
pinch of chilli powder
½tsp turmeric
salt
2tsp lemon juice
For the garnish:
1 medium-sized onion, sliced in rings, fried in a little oil and drained on absorbent paper
flat-leaved parsley, chopped
lemon wedges

- ***Serves 4*** 
- ***325cals/1365kjs per serving***

1 Cover the brisket with salted water and boil for 1 hour, skimming the surface from time to time to remove any scum. Meanwhile, chop and finely mince together the prawns, shallots and garlic.

2 Drain the meat, allow to cool slightly, and cut into bite-sized cubes, reserving the stock. Sauté the prawn mixture for 1 minute in the vegetable oil. Add the ginger, chilli powder, turmeric, salt and 225ml/8fl oz of the stock, cover and simmer for 7–8 minutes.

3 Put the meat cubes in another saucepan, strain over the prawn mixture and discard the solids left in the strainer. Simmer the meat for 2 minutes before adding the rest of the stock. Bring the soup to the boil and simmer for a further 20 minutes.

4 Add the lemon juice and serve the soup hot, garnished with the fried onion rings, flat-leaved parsley and lemon wedges in heated bowls.

Prawn and vermicelli soup

- ***Preparation: 30 minutes***
- ***Cooking: 1 hour 25 minutes***

100g/4oz pork fillet or chicken breast
175g/6oz raw prawns, shelled
1½tbls vegetable oil
5 shallots, finely sliced
2 garlic cloves, crushed
1tsp ground ginger
1tsp ground coriander
½tsp turmeric
175g/6oz rice vermicelli (laksa)
850ml/1½pt boiling water
300ml/10fl oz thick coconut milk (see page 140)
4 yellow bean curds, cut into thick strips
75g/3oz beansprouts
salt and freshly ground black pepper
2 spring onions, thinly sliced, to garnish

- ***Serves 6*** 
- ***315cals/1325kjs per serving***

1 Put 600ml/1pt water in a saucepan, season with salt and pepper, add the pork or chicken, bring to the boil and simmer for 45 minutes. Remove the meat with a slotted spoon and chop into small pieces. Reserve the stock.

2 Heat the oil in a wok or a deep saucepan over high heat, add the shallots and sauté for 1 minute. Add the ▶

garlic, ginger, coriander and turmeric, stirring for another 30 seconds before adding the prawns and the meat. Fry for a further 1 minute then add the stock and simmer for 25 minutes.

3 Place the vermicelli in a saucepan, pour over the boiling water, cover and leave for 5 minutes. Drain the vermicelli.

4 Add the vermicelli and the coconut milk to the wok and continue simmering very gently for 15–20 minutes until the coconut milk starts to boil. Stir well, add the bean curd strips and beansprouts and simmer for 5 minutes, stirring from time to time to avoid sticking.

5 Pour the soup into warmed bowls, garnish with the thinly sliced spring onions and serve immediately.

Crab with chilli

This dish is better if you only use the firm white flesh of the crab, so if using whole crabs, keep the soft brown meat for another recipe. If you can buy crab claws alone, they are ideal.

• **Preparation: 40 minutes**

• **Cooking: 12 minutes**

2 medium-sized crabs, boiled, or 8 large crab claws
3tbls vegetable oil
1tbls lemon juice
salt
For the sauce:
5 red chillies, seeded and chopped
1 onion, finely chopped
2 garlic cloves
1tsp powdered ginger
5tbls vegetable oil
2 ripe tomatoes, blanched, skinned, seeded and chopped, or 2tsp tomato purée
1tsp sugar (optional)
1tbls light soy sauce
salt

• **Serves 4** 

• **405cals/1700kjs per serving**

1 If using whole crabs, clean them, removing and discarding the dead men's fingers and removing the soft brown flesh which can be used in another recipe. Leave the white flesh in the shells.

2 Cut the bodies of the crabs into 4 pieces and chop the claws, if they are very large, into 2 or 3 pieces.

3 Heat 3tbls oil in a large frying-pan over medium-high heat, add the crab pieces and fry them for 5 minutes, stirring all the time.

4 Season with lemon juice and salt, then remove the crab with a slotted spoon and keep warm.

5 To make the sauce, pound the chillies, onions and garlic using a food processor and mix in the ginger.

6 Heat the oil in a wok over medium-high heat and stir-fry the chilli paste for 1 minute. Add the chopped tomato flesh, sugar, if using, and soy sauce. Stir for a further 2 minutes, adding some water if the paste is too thick and salt.

7 Simmer for a further 1 minute then add the crab pieces, stirring to coat with the sauce before serving.

Beef saté with peanut sauce

• **Preparation: 20 minutes plus 2 hours marinating**

• **Cooking: 20 minutes**

650g/1½lb rump steak, 1cm/½in thick
½tsp chilli powder
juice of ½ lemon
2tsp brown sugar
1tsp salt
1tsp ground coriander or cumin
For the peanut sauce:
2tbls vegetable oil
50g/2oz raw, shelled peanuts
2 red chillies, seeded and chopped or 1tsp chilli powder
2 shallots, chopped
1 garlic clove
1 slice balachan (shrimp paste)
1tsp brown sugar
juice of ½ lime
salt

• **Serves 4** 

• **285cals/1195kjs per serving**

1 Cut the steak into 1cm/½in cubes and put into a large bowl with the chilli powder, lemon juice, brown sugar, salt and ground coriander or cumin. Mix thoroughly and leave to marinate in a plastic bag for at least 2 hours, turn from time to time.

2 Meanwhile, make the sauce. Put 1tbls oil in a large frying-pan over medium-high heat, add the peanuts and stir-fry for 2–3 minutes. Remove from the pan and drain on absorbent paper.

3 Pound the chillies, shallots, garlic and balachan in a mortar and pestle until it is a smooth paste, or blend together.

4 Grind the peanuts to a fine powder. Put the remaining oil in a frying-pan over medium-high heat, add the chilli paste and fry for 1–2 minutes, then add 175ml/6fl oz water. Bring to the boil, add the ground peanuts, brown sugar, lime juice and a pinch of salt, and stir over a medium heat until the sauce is thick, about

10 minutes. Keep warm in a bowl. Heat the grill to high.

5 Thread the beef cubes onto bamboo skewers, 5 cubes per skewer, and grill for 10 minutes or until the meat is done, turning the skewers several times. Serve with the warm peanut sauce handed round separately.

Indonesian stuffed marrow or courgette

This recipe is from Sulawasi, the large island that used to be known as Celebes. It uses a *petola*, a young loofah that has not yet developed the fibrous inner skeleton sold as a bath sponge..

- ***Preparation: 30 minutes***
- ***Cooking: 1¼ hours***

2 small vegetable marrows or large courgettes
For the stuffing:
4 eggs
salt and freshly ground black pepper
vegetable oil for greasing
1tbls vegetable oil
5 shallots or 1 small onion, thinly sliced
2 green chillies, seeded and sliced into thin rounds, or ½tsp chilli powder
225g/8oz minced beef or chicken breast, minced
100g/4oz peeled prawns, chopped

- ***Serves 4***
- ***285cals/1195kjs per serving***

1 Heat the oven to 180C/350F/gas 4. If marrows are used, peel them thinly. Cut the marrows or courgettes down the middle lengthways, scoop out the seeds with a spoon or melon baller and discard them. Cover the marrow or courgette halves with cold, salted water.

2 To make the stuffing: reserve the white of one egg and beat the yolk with the rest of the eggs. Season with salt and pepper. Grease a 20cm/8in frying-pan with a little vegetable oil, then heat the pan. Pour enough of the beaten eggs in to make a thin omelette and cook over medium heat for 3–4 minutes.

3 Turn on to a plate to cool. Repeat until all the beaten eggs are used, greasing the pan with vegetable oil before cooking each omelette, and leaving the cooked omelettes to cool while you prepare the rest of the stuffing.

4 Sauté the shallots and chillies or chilli powder in 1tbls vegetable oil over low heat until they are soft. Mix the meat with the prawns, add them to the shallots and chillies and stir over medium heat for 3–4 minutes. Season the mixture with salt and pepper, remove from the heat and cool for 3–4 minutes, then add the reserved beaten egg white and mix well.

5 Remove the marrow or courgette halves from the water and dry them with absorbent paper. Roll up the omelettes, cut them into thin slices and mix them gently into the meat mixture. Fill two of the marrow or courgette halves with the mixture and cover with the other marrow or courgette halves. Place the stuffed vegetables on a baking sheet and bake in the over for 40–50 minutes. Serve cold, cut into thick slices.

Pancakes with coconut filling

The coconut filling of these sweet pancakes is characteristically Indonesian. If you use fresh coconut, remember to cut away the brown rind after you have prised the flesh out of the shell.

- ***Preparation: 20 minutes plus 2 hours standing***
- ***Cooking: 30 minutes***

150g/5oz flour
¾tsp salt
3 eggs
3tbls melted butter or vegetable oil
225ml/8fl oz milk
clarified butter or vegetable oil for greasing
For the filling:
75g/3oz brown sugar
100g/4oz desiccated coconut or ½ fresh young coconut, grated
1 cinnamon stick
pinch of salt
1tsp lemon juice

- ***Makes approx 15 pancakes***
- ***130cals/545kjs per pancake***

1 Sift the flour and salt into a bowl. Beat the eggs and stir them into the flour with the melted butter or oil. Stir gently until smooth. Gradually add the milk. Strain the batter through a fine sieve and dilute with a little water, if necessary, to give the consistency of thin cream. Leave the batter to stand for 2 hours.

2 Make the filling. Bring 225ml/8fl oz water to the boil in a saucepan, add the sugar and stir until it is dissolved. Then add the coconut, cinnamon stick and salt. Simmer the mixture gently over low heat until the coconut has absorbed all the water. Put in the lemon juice, stir for 1 minute, then discard the cinnamon stick. Keep the coconut filling warm while making the pancakes.

3 To make the pancakes, heat a 12–18cm/5–7in pan and grease it with a little clarified butter or vegetable oil. Spoon in just enough of the pancake batter to thinly coat the bottom of the pan, turning the pan to distribute the batter.

4 Cook the pancake for 1 minute over low heat and carefully turn it over using a flexible spatula. Cook the pancake for 1 minute on the other side.

5 Repeat until all the pancake batter is used, greasing the pan before cooking each pancake. When they are ready, fill each pancake. Fill each one with 1tbls of filling and roll before serving.

Far Flung Corners

The influences of France and China meet in food from this far corner of the world

Thai beef curry (page 86) with Fried mange-tout (page 85)

THROUGHOUT SOUTH-EAST Asia, food is noted for its combinations of hot spices and delicate flavours as well as for its exquisite presentation. It is usually as good to look at as to eat. In both Thailand and Indochina, food is served in bowls which come to the table all at the same time. Rice forms the basis of all meals and delicious dipping sauces add extra flavour and fragrance to the meal.

Thai delights

Cooking in this country blends Chinese stir-fry technique with the forcefulness of Indian curry spices: the result is unmistakable and unique in the Far East. Its special flavours include the famous fish sauce, *nam pla*, and sharp lemon grass, as well as the entire coriander plant, leaves, stems, roots and seeds. Ginger is also a favourite flavouring and is a distinctive element in Thai curries. Food can be amazingly hot, thanks to the generous use of dried and fresh chillies, and coconut milk adds its distinctive fragrance.

Around the Mekong delta

Here lie Laos, Cambodia and Vietnam, once part of France's overseas empire and therefore much influenced by French cooking. Conversely, the Vietnamese habit of wrapping food in edible leaves such as lettuce has undoubtedly had its effect on French nouvelle cuisine! Here too fish sauce – under the name of *nuoc nam* – appears in soups and sauces and the flavours of lemon grass, coriander and coconut milk predominate. Two sorts of rice are used – long-grain in Vietnam and Cambodia and glutinous short-grain in Laos. This sticky sort of rice is rolled into balls and used as a pusher, rather as the French use bread, throughout the meal.

Busy market selling vegetables and spices in Bangkok

Ginger beef soup

- ***Preparation:*** ***1¼ hours, plus 30 minutes chilling***
- ***Cooking: 2½ hours***

700g/1½lb oxtail, cut into chunks
700g/1½lb stewing steak in one piece
1 garlic clove, chopped
6 shallots, or 2 medium-sized onions, sliced
5cm/2in piece fresh ginger root, peeled
1 star anise
green part of 6 spring onions, chopped
4tbls chopped fresh coriander leaves
1 large onion, halved, then very thinly sliced
125g/4oz been sprouts
225g/8oz topside beef in one slice
225g/8oz rice stick noodles
*2tbls Thai fish sauce (*nam pla*)*
1tsp salt
pinch of monosodium glutamate
1 lime or lemon cut into 8 wedges
1-2 fresh hot red chillies, seeded and sliced

- ***Serves 8*** 
- ***465cals/1955kjs per serving***

1 Put the oxtail, garlic, shallots or onions, ginger root and star anise in a large saucepan with 2L/3½pt cold water. Bring the liquid to a boil over medium heat, skim, lower the heat and simmer, covered, for 1 hour. Add the stewing steak and continue to simmer over a low heat until the meats are tender, which will take about 2 hours.

2 Meanwhile, combine the chopped spring onions with the coriander leaves in a small bowl. Put the sliced onion into another bowl. Drop the bean sprouts into boiling water for 30 seconds, drain, rinse in cold water and put them into another bowl. Put the topside of beef into the freezer to chill for 30 minutes.

3 Bring a large saucepan of water to a boil and add the rice stick noodles. Bring the water back to a boil over high heat and boil for 3 minutes. Drain the noodles in a colander, rinse in cold water, then drain thoroughly. Cut the chilled beef into the thinnest possible slices and reserve them in a bowl.

4 Lift the cooked meats out of the soup and, when they are cool enough to handle, remove and discard the bones from the oxtail and slice the chuck or stewing meat. Add the fish sauce, salt and monosodium glutamate to the soup. Return the prepared meats to the soup and heat slowly but thoroughly.

5 Have a large bowl prepared and heated for each guest. Divide the noodles between the bowls, top with the cooked meats, then add the thin slices of raw beef. Add about half of the onion slices, then some bean sprouts and a little of the mixed coriander and spring onion green. Have the soup ready and boiling hot; fill up each bowl; the heat from the soup will cook the raw beef.

6 At the table, guests add more bean sprouts and onion and squeeze lime or lemon juice into the soup if wished. The chopped fresh chillies and extra fish sauce are served separately, and can be added to suit individual tastes.

Fried noodle salad

This is an extremely popular luncheon dish in Thailand. It can be accompanied by curry or a fish or shellfish dish, or it may be served as the main course with soup and dessert.

- ***Preparation: 1 hour***
- ***Cooking: 30 minutes***

225g/8oz Chinese rice flour noodles
1tbls sugar
1tbls soy sauce
1tbls distilled or white wine vinegar
vegetable oil for deep frying
50g/2oz shallots, finely chopped
4 garlic cloves, finely chopped
250g/9oz boneless pork loin or

250g/9oz boneless raw chicken breast, cut into thin strips
250g/9oz flaked crabmeat or shelled prawns, cut into 1.5cm/½in pieces
2tbls fish sauce (nam pla, *page 86*)
2tbls lemon juice
4-egg omelette, cut into strips
6 fresh red or green chillies, slit and seeded and then slit in 4-5 places from tip almost to stem end to make 'flowers'
6 spring onions, cut into 5cm/2in pieces
coriander sprigs
225g/8oz bean sprouts

• Serves 6 

• 415cals/1745kjs per serving

1 Cover the noodles with hot water and let them stand for 1 minute. Drain and let them dry. Mix together the sugar, soy sauce and vinegar and reserve.

2 Heat about 5cm/2in vegetable oil in a wok or deep-fat frier to 220C/425F or until a cube of bread will brown in 30 seconds. Fry the noodles, a handful at a time, until they are golden, turning once. Lift them out and drain on absorbent paper. The noodles will break up.

3 Heat 3tbls vegetable oil in a wok or frying-pan over medium heat. When hot, stir-fry the shallots and garlic for 30 seconds. Add the pork or chicken breast and stir-fry for 2 minutes. Add the crab-meat or prawns and stir-fry for 1 minute longer. Add the *nam pla* and lemon juice, stir to mix and cook for 1 minute.

4 Toss the noodles into the meat mixture and pour over the reserved vinegar mixture. Cook over moderate heat just long enough to heat through.

5 Turn the noodle mixture on to a large warmed platter and garnish it with the omelette strips, chilli 'flowers', spring onions, coriander sprigs and bean sprouts. Serve immediately.

Cook's tips

Uncooked prawns and crab meat give the best flavour to this dish, but frozen will do.

Spicy prawn soup

This classic Thai soup is easy to make; if lemon grass, often sold ground as *serehpoeder* or *sereh* powder, is not available, use a piece of lemon zest.

• Preparation: 15 minutes

• Cooking: 18 minutes

1.5L/2½pt light chicken stock
1tsp ground lemon grass or 25mm/1in square piece of lemon zest
2tbls lemon juice
75ml/3fl oz fish sauce (nam pla)
1-2 fresh hot chillies, seeded and sliced (optional)
3tbls coarsely chopped coriander leaves
6 spring onions, trimmed and sliced across thinly
250g/9oz frozen prawns, defrosted

• Serves 6 

• 55cals/230kjs per serving

1 Combine the chicken stock, lemon grass or zest, lemon juice, *nam pla* and chillies in a saucepan. Bring to the boil, cover and simmer over low heat for 15 minutes stirring occasionally.

2 Strain the mixture through a fine sieve into another saucepan. Discard the solids collected in the sieve.

3 Add the coriander leaves, spring onions and prawns to the stock and simmer over low heat for 2 minutes or until the prawns are properly heated through without boiling which would toughen the prawns. Serve immediately.

Fried mange-tout

This is a favourite Thai vegetable recipe. Mange-tout peas are eaten pod and all; choose the smallest, youngest available. Cabbage or Chinese leaves can be used instead of the mange-tout.

• Preparation: 15 minutes

• Cooking: 15 minutes

2tbls vegetable oil or lard
1tbls finely chopped garlic, or less to taste
50g/2oz lean raw pork, chopped
25g/1oz boiled, peeled prawns, chopped
350g/12oz mange-tout, stem end removed
1tbls fish sauce (nam pla)
1tsp sugar

freshly ground black pepper

- ***Serves 4*** 
- ***130cals/545kjs per serving***

1 Heat the oil in a frying-pan over medium-low heat and sauté the garlic until it is lightly coloured. Add the pork and sauté until it is lightly browned.

2 Add the prawns and mange-tout and sauté the mixture, stirring, until the peas are cooked, but still crisp, about 3 minutes. Stir in the *nam pla* and sugar. Season with pepper to taste and serve immediately.

RED CURRY PASTE

This curry paste will keep up to 1 month, refrigerated in a screw-top glass jar. Use for poultry or beef curry (see below) or to flavour mayonnaise. 15g/½oz dried hot chillies, seeded; 15g/½oz dried shrimps; 3 shallots, chopped; 1tbls chopped fresh root ginger; 1tbls chopped garlic, 1tsp black peppercorns; 1tsp ground coriander; 1tsp ground cumin; 1tbls chopped fresh coriander, using the leaves, stems and roots; ½tsp ground *laos* or galingale (see right); 1tsp ground lemon grass or 1tbls chopped lemon zest; 2tbls lemon juice. Soak chillies and shrimps in hot water to cover until soft, about 20 minutes. In a blender, combine the chillies and shrimps and any soaking water with the remaining ingredients. Blend to a thick paste. Add extra lemon juice if needed. This makes 175g/6oz.

Thai beef curry

Thai beef curry with its contrasting flavours is deliciously different from the curries of other countries. It can be served as an accompaniment to Fried noodle salad (see recipe), or with rice and a vegetable dish.

- ***Preparation: 1 hour***
- ***Cooking: 1¼ hours***

400ml/14fl oz thin coconut milk (see page 140)
450g/1lb lean beef cut into 25 x 15mm/1 x ½in strips
2 tbls Red curry paste (see left)
½tsp ground cardamom
¼tsp ground mace
1tbls fish sauce (nam pla)
50g/2oz coarsely chopped basil leaves or 1tbls dried basil soaked in warm water for 5 minutes and drained

- ***Serves 4*** 
- ***300cals/1260kjs per serving***

1 In a saucepan bring the coconut milk to a boil. Add the beef strips, reduce the heat to low and simmer, partly covered, for 1 hour, or until the beef is tender. Alternatively cook in the oven at 180C/350F/gas 4 until tender.

2 Stir in the curry paste, cardamom, mace and *nam pla* and simmer the curry uncovered for 15 minutes longer. Just before serving, stir in the basil.

Cambodian fruit salad

This simple and refreshing fruit salad gets its unusual taste from lichees.

- ***Preparation: 35 minutes, plus 2-3 hours chilling***

1 medium-sized grapefruit, peeled and sectioned
1 medium-sized orange, peeled and sectioned
1 small pineapple, peeled, cored and cubed, or 450g/1lb canned pineapple cubes, drained
14 fresh lichees, peeled, stoned and halved, or 450g/1lb canned lichees, drained and halved, reserving juice
1tbls sugar (optional)
1tsp lemon or lime juice

- ***Serves 4*** 
- ***105cals/440kjs per serving***

1 Remove any seeds and white pith from the grapefruit and orange sections and cut them across into halves.

2 In a glass dessert bowl combine the orange and grapefruit sections with the pineapple and lichees. Add 2tbls of the reserved lichee syrup (if using canned fruit), or sprinkle with the sugar. Add the lemon or lime juice and mix lightly.

3 Chill for at least 2 hours before serving in individual dishes.

SPECIAL FLAVOURS

Fish gravy, fish sauce, fish soy, *nam pla, nuoc nam* . . . this is made from fresh anchovies which are layered with salt and allowed to ferment in barrels. It smells rather astonishing to the Western nose. *Laos* or galingale looks rather like ginger root but is quite different in flavour. If you cannot buy it, leave it out, as there is no substitute.

Baltic to Bohemia

Cold winters have influenced the cooking of Central Europe with its tradition of hearty soups and hotpots

Fried carp in breadcrumbs (page 88)

THERE ARE MANY similarities between the cooking of Poland and Czechoslovakia, partly because both countries have a wealth of forests, which has brought a passion for mushrooms rivalled only by that of their Russian neighbours, and partly because the extreme bitterness of the winter weather demands the kind of cooking that keeps out the cold.

Winter warmers

One of the great casseroles of the world is Poland's *bigos*, a long-simmered blend of spicy sausage, raw ham, apples and, most important, fresh cabbage and sauerkraut. It was originally cooked to be taken on long hunting expeditions. Other favourite winter soups include *krupnik*, heavy with barley, and *lapusniak*, which is cabbage-based. Most famous of all is beetroot soup, served cold in summer (see page 91) but hot in winter, particularly on Christmas Eve. Dumplings, too, add body to soups, particularly in Czechoslovakia where they are made with breadcrumbs.

Mushroom mania

The woods and forests of these two countries still produce large quantities of edible fungi – best of all is the cep, *Boletus edulis*, which has the advantage of drying well so that it can appear in Polish and Czech cooking throughout the winter. Poles call these 'real mushrooms' and their flavour is much stronger and delicious than the store-bought button variety. Czechs like to grill them fresh with marjoram and cayenne pepper.

Olden days

There was a time when the forests were filled with wild boar, but the days of the great boar-hunt are long gone. There is still plenty of other game and indeed ordinary meat is often cooked with a juniper and peppercorn marinade which makes it taste very much like a traditional game dish (see page 89). Pork is a real favourite in Central Europe – and caraway seeds are a traditional spicy flavouring. Geese are still top favourite for feasts and special occasions, again spiced with caraway seeds, and goose fat is carefully kept to use in preliminary frying. Old cooking traditions die hard in this part of the world and traditional recipes are still used when the proper ingredients are available.

Red and green peppers for sale on a local market stall

Fried carp in breadcrumbs

- ***Preparation: 25 minutes plus 30 minutes standing time***
- ***Cooking: 30 minutes***

4 carp steaks, 2.5cm/1in thick
salt
juice of 1 lemon
flour for coating
1 large egg, beaten
1tsp grated nutmeg
fresh white breadcrumbs, for coating
100g/4oz butter
mixed salad, to serve

- ***Serves 4***
- ***430cals/1805kjs per serving***

1 Rinse the carp steaks quickly in cold water and pat them dry with absorbent paper. Lightly salt them, pour over the lemon juice, cover and leave to stand for 30 minutes.

2 Wipe the carp and salt them again. Put the flour on a plate, the beaten egg and the nutmeg into a shallow dish, and the breadcrumbs, onto a plate. Dip each steak into the flour, shake, then dip into the egg and the breadcrumbs.

3 Melt half the butter in a large frying-pan over a moderate heat, and fry the steaks for 5 minutes on each side. Heat the oven to 170°C/325°F/gas 3.

4 Melt the remaining butter in a small saucepan, pour it into a shallow ovenproof dish, put in the carp steaks and bake for 15 minutes. Serve at once with a mixed salad.

FRESH FISH

Poland has the advantage of its Baltic coastline, so makes full use of its plentiful herring catches. Most of the dishes are the same as those from the Baltic republics. The herring are most often pickled or salted and served with a glass of vodka as an appetizer. Czechoslovakia is landlocked so the fish it produces are the freshwater variety, carp being one of the top favourites.

Bohemian chicken

You can serve the sauce separately as a soup before eating the delicately flavoured chicken as a main course.

- ***Preparation: 40 minutes if jointing chicken***

◀ • ***Cooking: 1¾ hours***

1.5kg/3lb chicken, cut into 6 joints
salt
1 carrot, chopped
2tbls parsley, finely chopped
1 Spanish onion, finely chopped
1 leek, finely sliced into rings
freshly ground black pepper
2 tbls flour
25g/1oz butter
freshly grated nutmeg
1 small cauliflower, cut into florets
noodles or rice, to serve

• ***Serves 6*** 

• ***475cals/1995kjs per serving***

1 Skin the chicken pieces and put them into a large saucepan. Cover with about 2.3L/4pt water, lightly salted, bring to the boil, then simmer for 1 hour.

2 Add the carrot, parsley, onion and leek and cook for a further 30 minutes.

3 Season the mixture with salt and freshly ground black pepper. Make a beurre manié by mixing the flour and butter together with a fork, then add the mixture to the pan, stirring well till the sauce thickens slightly. Add nutmeg to taste, then simmer.

4 Bring a saucepan of water to the boil, add salt and drop in the cauliflower florets. Cook until tender but not mushy, about 5-7 minutes. Drain and add the cauliflower to the chicken mixture, stirring gently but thoroughly.

5 Check for seasoning and serve immediately with noodles or rice.

Pot roast

Veal, lamb or mutton can also be cooked in this way. If lean, the meat should be larded with pork fat before marinating. Dried Polish mushrooms (ceps) can be bought from Italian grocers.

• ***Preparation: 40 minutes plus 24 hours marinading***

• ***Cooking: 3 hours***

2kg/4½lb stewing beef, suitable for pot roast (top or back rib or similar, boned or unboned and rolled)

For the marinade

300ml/¼pt cider, wine or spiced vinegar
100ml/3½fl oz red wine (optional)
2.5cm/1in parsley root or a small bunch of parsley stalks
2 celery stalks, chopped
1 onion, chopped
20 black peppercorns
10 juniper berries
1 bay leaf
3 sprigs thyme or ½tsp dried thyme
salt

For the pot roast

50g/2oz pork or chicken fat, or dripping
1 large onion, diced
4 dried ceps or dried mushrooms, soaked
2 celery stalks, diced
2.5cm/1in parsley root or a small bunch of parsley stalks
3 sprigs fresh thyme or ½tsp dried thyme
1tbls flour
175ml/6fl oz soured cream

• ***Serves 8*** 

• ***550cals/2310kjs per serving*** ▶

◀ 1 To make the marinade, pour the liquid into a saucepan, together with 1L/1¾pt water, add the rest of the marinade ingredients, bring to the boil, and simmer for 30 minutes. Leave the liquid until it is completely cold. Place the meat in a large glass or earthenware bowl and pour over the marinade. Leave the meat in the marinade in a cool place for at least 24 hours, turning it at least once every 12 hours. If the meat is left to marinate in a very cold larder or in the refrigerator, it can be left for up to 8 days. The night before it is to be cooked, soak the dried mushrooms in water to cover.

2 To cook the meat, remove it from the marinade 1 hour before cooking and rub it all over with coarse salt. Strain the marinade and discard the liquid; reserve the vegetables for the pot roast. Strain and reserve the mushroom soaking liquid.

3 Wash the salt from the meat and pat it dry with absorbent paper. Heat the oven to 150°C/300°F/gas 2.

4 In a frying-pan, melt the fat or dripping and turn the meat in it until it is nicely browned on all sides. Transfer the meat to a casserole with a tight-fitting lid. Add the marinade vegetables, soaked ceps or mushrooms and the pot roast vegetables and spices. Cover the casserole and cook it in the oven for at least 3 hours, basting every half hour with 2tbls of the mushroom liquid mixed with cold water.

5 Mix the flour and soured cream and add them to the casserole. Cook the meat for a further 30 minutes.

6 To serve the meat, slice and arrange it in overlapping layers on a large warmed serving dish. Keep the meat in a warm place while the cooking liquid is strained, then pour it over the meat and serve hot.

Czech marinated beef

- ***Preparation: 45 minutes plus 4-6 days***
- ***Cooking: 2½ hours***

1kg/2¼lb fillet of beef, in one piece
100g/4oz fat from smoked streaky bacon
salt
3tbls vegetable oil
1tbls flour
500ml/18fl oz yoghurt
1-2tsp lemon juice (optional)
lingonberries or cranberries in syrup, to serve

For the marinade
½ Spanish onion, finely chopped
1 stick celery, finely chopped
1 carrot, chopped
2-3 sprigs of parsley
1 bay leaf
4 whole allspice berries
8 black peppercorns
pinch of thyme
5tbls red wine vinegar

- ***Serves 8*** (₶)(£££)
- ***335cals/1405kjs per serving***

1 Put all the ingredients for the marinade into a saucepan with 600ml/1pt water. Bring the liquid to the boil, simmer for 5 minutes, then cool.

2 Chop the fat into strips 2.5cm/1in long, 5mm/¼in wide, and insert the strips in the beef all over with a larding needle. Lightly salt the meat and put it into a deep casserole dish. Pour over the marinade and leave, covered, in the refrigerator for 4-6 days, turning the meat once every day.

3 Remove the meat from the marinade and take the vegetables out with a slotted spoon. Heat the oven to 150°C/300°F/gas 2.

4 Heat the oil in a large frying-pan over medium heat and brown the meat on all sides. Put the meat back in the casserole dish, with the marinade.

5 Fry the vegetables for 2 minutes in some oil and add them to the casserole. Put the casserole in the oven for 2-2½ hours, or until the meat is tender.

6 When the meat is done, remove it from the dish, slice it and keep warm.

7 Mix the flour with the yoghurt, stirring well to get rid of any lumps. Pour the juices and vegetables from the casserole dish into a large saucepan and add the yoghurt mixture. Bring the mixture to the boil, stirring, then remove from the heat and sieve it into a warmed sauceboat.

8 Taste the sauce and add lemon juice if necessary. Pour a little sauce over the beef and serve the dish immediately. Pour the rest of the sauce in a warm sauceboat and serve it and the lingonberries or cranberries in syrup separately.

Summer beetroot soup

This very healthy, refreshing starter, whose Polish name simply means 'cold soup', can also be garnished with watercress or the green tops of beetroot and with very thin slices of lemon. You can prepare the soup a day ahead, refrigerate and add the other ingredients and garnishes just before serving. For a light summer lunch, serve the soup as a main course with crusty bread. Using green beetroot tops instead of spinach makes a more authentic version of this dish.

- ***Preparation: 25 minutes***
- ***Cooking: 1¼ hours***

600g/1lb 6oz raw beetroot
400g/14oz spinach, trimmed
250ml/9fl oz veal or chicken stock
salt
50ml/2fl oz wine vinegar or lemon juice
250ml/9 fl oz soured cream
2 large cucumbers, peeled and chopped
6 spring onions, chopped
6 large radishes, thinly sliced
3tbls dill weed or fresh dill
large pinch of white pepper
For the garnish
400g/14oz boiled, peeled shrimps
2 eggs, hard-boiled

- ***Serves 8***
- ***195cals/820kjs per serving***

1 Wash the beetroot carefully without damaging the skin. Put them into a large saucepan and cover with water. Bring to the boil and simmer over low heat for 40-60 minutes until tender. Remove the beetroot from the pan with a slotted spoon, and set aside.

2 Bring the cooking liquid back to the boil, salt lightly and add the spinach. Boil the spinach rapidly for 2 minutes, drain it and reserve the cooking liquid in the pan.

3 Add the stock to the cooking liquid, season with more salt and simmer the liquid for 10 minutes. Meanwhile finely chop the spinach and half the beetroot (reserving the rest for another dish).

4 Remove the pan from the heat, cool and add the chopped spinach and beetroot. Chill the soup until very cold.

5 Just before serving, stir in the vinegar or lemon juice and the soured cream. Stir in the cucumbers, spring onions, radishes, dill weed or dill and pepper. Cut the hard-boiled eggs into 8 pieces lengthways. Pour the soup into chilled bowls and garnish with shrimps and hard boiled eggs.

Cook's tips

To serve the soup hot, leave out the shrimps and cucumbers and serve extra soured cream.

EASTER CAKES

Poland is one of the countries which uses a sweet yeast dough to make a traditional cake. Babka, or 'grandmother', is so-called because its mould looks like the wide skirts of a peasant woman; it is very like a *guglhopf* but at its Easter best it is enriched with almonds, raisins and candied peel.

Plum meringue slice

There are many ways of making this dessert. Recipes for the dough and topping can vary considerably, but the cake is always baked in a rectangular tin with sides about 2.5cm/1in high, usually lined with rice paper. Ground almonds or walnuts are usually included in the dough.

● ***Preparation: 25 minutes***

● ***Cooking: 1 hour***

225g/8oz butter
225g/8oz caster sugar
300g/11oz flour
5 egg yolks
vanilla essence
100g/4oz ground almonds
1tsp lemon zest, grated
pinch of salt
rice paper (optional)
For the topping
5 egg whites
75g/3oz caster sugar
pinch of salt
400g/14oz plum butter or any non-sweet stiff fruit purée

● ***Serves 12*** 

● ***495cals/2080kjs per serving***

1 Heat the oven to 170°C/325°F/gas 3. Cream the butter and caster sugar together until smooth. Then beat 1tbls of the flour alternately with each of the egg yolks. Add a few drops of vanilla essence to taste, the ground almonds, lemon zest and salt and mix in the rest of the flour.

2 Line a 38 x 50cm/15 x 20in Swiss roll tin with rice paper, if using, and butter the sides. Roll out or pat the dough into shape and fill the tin. Smooth the dough with a palette knife so that it fits exactly.

3 Bake the dough in the oven for 35-40 minutes or until it is golden brown. Meanwhile for the topping, whip the egg whites with the salt until they peak. Whip in the sugar until stiff peaks form.

4 Remove the dough from the oven and raise the heat to 190°C/375°F/gas 5. Spread the plum butter on the cake while it is still warm to allow for easy spreading. Pile the whipped meringue mixture on top of the jam and return the cake to the oven. Bake it for 10 minutes or until the meringue is set and is just beginning to brown. Cut into slices and serve warm.

Cherry sponge

● ***Preparation: 25 minutes***

● ***Cooking: 35 minutes***

150g/5oz butter, softened
150g/5oz caster sugar
3 large eggs, separated
150g/5oz flour, sifted
400g/14oz fresh black cherries or drained, canned black cherries, stoned
icing sugar, to dust
butter for greasing

● ***Serves 8*** 

● ***345cals/1450kjs per serving***

1 Heat the oven to 180°C/350°F/gas 4. Line a 8 x 8in/20 x 20cm roasting tin with buttered paper.

2 Mix the butter and sugar together until they are light and creamy. Add the egg yolks gradually and beat thoroughly.

3 Whisk the egg whites until they form soft peaks and add them alternately with spoonfuls of the flour folding into the egg, butter and sugar mixture. Spread the sponge mixture evenly over the tin.

4 Wash the cherries or drain them well on absorbent paper, if you are using canned ones. Lay them thickly over the sponge mixture. Bake for 30-35 minutes or until a skewer pierced into the sponge comes out clean. Cool, dust with icing sugar, cut into slices and serve.

Balkan Blends

A changing blend of peoples and cultures gives Balkan food an unequalled variety of flavours

Vegetable casserole (page 94)

THE BALKANS HAVE always been a meeting ground for different peoples and, as boundaries and borders have changed with the tide of history, Bulgaria, Yugoslavia and Romania have absorbed culinary traditions from all over Central Europe.

Romanian recipes

In the early part of this century, Romania was famous for the richness and variety of its agricultural produce and for this reason many of the great Romanian dishes are fruit- and vegetable-based. *Ghiveciu* is a delicious vegetable stew found all over the Balkans – a varying combination of peppers, aubergines, tomatoes, okra and onions, but in Romania it has an unusual tart savour because it includes unripe seedless grapes, greengages and gooseberries (see page 94). A similar sourness is typical of the famous Romanian soup, *ciorbă*, which includes unripe grape juice, unripe fruit, vinegar or lemon.

Slav selections

Yugoslavia is a hotch-potch of different provinces and its cooking is an astonishing blend of traditions from the Middle East, Central Europe and the warm Mediterranean. No wonder that its typical dish, Vegetable macédoine (page 96), is named after the mixture of people found in Macedonia. The strong Muslim influence in Bosnia means that lamb is a favourite meat hereabouts. In Slovenia, which is basically Roman Catholic, the typically Central European combinations of pork and cabbage are a favourite.

Bulgarian yoghurt

Bulgarians seem to have been passionate yoghurt-eaters since time immemorial and this is apparently because the bacteria which turn all types of milk into yoghurt occur naturally in this country. Cheese too is a great favourite, mainly crumbly white Sirene which is rather like the Greek feta. Like the other Balkan countries, Bulgarian cooking makes extensive use of tomatoes and peppers and it is maize rather than wheat which provides flour for their daily bread.

Large-scale goulash preparation in a Balkan kitchen.

Romanian vegetable and fruit casserole

- ***Preparation: 2 hours, plus 1 hour standing***
- ***Cooking: 2¾ hours***

1 aubergine, in 25mm/1in cubes
salt and pepper
200g/7oz small French beans, fresh or frozen and defrosted, trimmed and cut into bite-sized pieces
150g/5oz shelled green peas, fresh or frozen and defrosted
6tbls vegetable oil; extra for greasing
2 onions, finely chopped
250g/9oz courgettes, cubed
1 large green or red pepper, seeded and cut into 2cm/¾in squares
400g/14oz potatoes, in 1.5cm/½in cubes
1 carrot, peeled and diced
100g/4oz celeriac, peeled and cut into 1.5cm/½in cubes
1tsp paprika
100g/4oz tender okra pods, no more than 5cm/2in long
250ml/9oz tomatoes, blanched, peeled and chopped, or canned tomatoes, chopped
4 garlic cloves, peeled and finely chopped
100g/4oz tart (unripe) seedless green grapes, tart (green) greengage plums, halved and stoned, or tart gooseberries
3tbls finely chopped parsley
3-4 tomatoes, sliced into rounds
cucumber salad and fresh bread, to serve

- ***Serves 4*** 
- ***380cals/1595kjs per serving***

1 Put the cubed aubergine in a bowl, sprinkle with 1tsp salt and allow to stand for 1 hour.

2 Meanwhile, if using fresh beans and/or peas, put them in a saucepan with 4tbls water, covered, over low heat to steam for about 15 minutes, or until they are partially cooked and the liquid has evaporated. Remove from the heat.

3 Heat the oven to 200C/400F/gas 6 and place a shelf just above the centre of the oven. Oil the inside of a wide shallow earthenware casserole, about 30cm/12in in diameter and about 9cm/3½in deep. Rinse and drain the aubergine, then squeeze to press out the bitter juices. Put the aubergine into the casserole.

4 Add the cooked or frozen beans and peas, the onion, courgettes, green or red pepper, potatoes, carrots and celeriac to the casserole. Season with the paprika and freshly ground black pepper and pour in 5tbls of the oil. Mix everything together with your hands, then lightly press the mixture down into the casserole. Bake, uncovered, for 1 hour.

5 Remove the casserole from the oven and stir in the okra, tomatoes, garlic, fruit and 2tbls parsley. Smooth the surface with the back of a spoon and arrange the sliced tomatoes on top. Sprinkle with the remaining oil, then bake a further 1¾ hours, or until the tomato slices are lightly browned and the liquid has evaporated, giving the casserole a rich glaze.

6 Sprinkle with the rest of the parsley 5 minutes before cooking time is up and finish cooking. Cool and serve in the casserole, with a cucumber salad and fresh bread.

STUFFED VEGETABLES

In Bulgaria, as in Greece and Turkey, stuffed vine leaves are a favourite dish: the stuffing is a tasty blend of rice, minced lamb and herbs, but there is a difference – the leaves are served with a subtle yoghurt-based sauce thickened with egg yolks. Accompany with a bowl of chilled yoghurt.

Grilled trout with clotted cream

- ***Preparation: 15 minutes, plus filleting time and 2 hours marinating***
- ***Cooking: 20 minutes***

2 × 350g/12oz brown or rainbow trout, skinned and filleted
juice of 1 lemon
100g/4oz clotted cream
salt and freshly ground black pepper
15g/½oz butter, softened
For the garnish:
250g/9oz frozen peas
250g/9oz button mushrooms, trimmed
25g/1oz butter
4lemon slices

- ***Serves 4*** 
- ***420cals/1765kjs per serving***

1 Place the fillets in a close-fitting dish and pour over the lemon juice. Marinate, covered, for 2 hours in a refrigerator.

2 Season the cream with salt and pepper to taste and spread it on the base of a shallow rectangular ovenproof dish large enough to hold the fillets snugly in a single layer. Lay the fillets over the cream, season to taste and dot them evenly with the butter.

3 Remove the grid from a grill pan and heat the grill to medium high. When hot, put the prepared dish into the grill pan and grill for 10 minutes at high heat, then turn down the grill to medium low and cook for a further 5 minutes, or until the fish flakes with fork.

4 Meanwhile, cook the peas in boiling salted water for 5 minutes and drain. Sauté the whole mushrooms in butter for 2-3 minutes.

5 Using a fish-slice or a long spatula, transfer the fillets, without breaking them, to a round, well-heated serving dish, laying them side by side. Spoon a row of the peas on one side of the trout, and place a row of the mushrooms on the other side. Pour the pan juices from the ovenproof dish over the fillets, garnishing them with the lemon slices. Serve immediately.

Belgrade bite-sized sausages

- ***Preparation: 30 minutes, plus chilling time***
- ***Cooking: 15 minutes***

900g/2lb boneless beef: back rib, skirt, flank or neck, or a mixture of these (butcher's mince is unsuitable)
200g/7oz fresh beef suet, or raw beef marrow
1tsp salt
½tsp freshly ground black pepper
1tbls finely chopped onion (optional)
oil for greasing
For the garnish:
2 onions, finely chopped
salt
6-8 large pickled pimentos or gherkins
6-8 fresh whole chillies, pricked with a fork and grilled (optional)

- ***Serves 6*** 
- ***385cals/1615kjs per serving***

1 Trim all connective tissue and membrane from the meat and the suet. Chop them very finely, first separately and ▶

then together with heavy sharp knives until a perfectly smooth meat pulp is produced; or purée them together in a food processor.

2 Combine the beef purée in a bowl with the rest of the sausage ingredients and knead thoroughly. Fry a teaspoonful of the mixture, then taste and adjust the seasoning if necessary. Cover and reserve for 30 minutes, or refrigerate, lightly covered, for up to 12 hours.

3 With dampened hands, shape the sausage meat into 60 walnut-sized balls Put each meat-ball on a work surface and with the palm of your hand, roll it backwards and forwards until it is about 5cm/2in long. Straighten up the ends by tapping them with your finger-tips. Refrigerate the sausages until you are ready to cook them. Rub the grid with oil and place sausages on grid. Turn up the heat of the grill to high.

4 Grill the sausages, initially close to the heat source, searing them at high heat for 1 minute on each side. Then reduce the heat and cook them for a further 8-10 minutes, or until well coloured on the surface and still succulent inside.

5 Sprinkle the chopped onions with salt, and serve the sausages immediately, garnished with the onions, pickled pimentos or gherkins and the whole grilled chillies, if wished.

Croatian vegetable macédoine

- ***Preparation: 20 minutes***
- ***Cooking: 25 minutes***

25g/1oz butter
3 medium-sized carrots, in 5mm/ 1/4in dice
2 medium-sized potatoes, in 1.5cm/ 1/2in cubes
1 large head celeriac, cut in 1.5cm/ 1/2in cubes
100g/4oz frozen peas
1tsp salt
freshly ground black pepper
roast beef to serve

- ***Serves 4***
- ***135cals/565kjs per serving***

1 Melt the butter in a large frying-pan over low heat. Add the diced carrots, potatoes, celeriac, salt and pepper to taste. Stir the vegetables together until they are well mixed and glistening with the butter.

2 Put on a tightly fitting lid and let the vegetables sweat over the lowest possible heat, without browning, for 15 minutes, add the peas. Cook for a further 5 minutes or until the vegetables are just soft enough to eat but have not lost all their firmness. Taste for seasoning and spoon them into an oven-to-table serving dish. Keep warm.

3 Pour off the fat from the pan in which the beef has been roasted. Discard the fat. Drizzle the beef pan-juices over the vegetables. Keep the dish hot until serving.

SUNFLOWER SPLENDOUR

All over the Balkans there are great fields of sunflowers and it is the oil they produce that is mostly used in cooking. Butter is very much a rarity, but a small amount of it appears in vegetable macédoine. In Croatia, where pork is popular, lard is used for frying.

Baked cornmeal with cheese or bacon

The Bulgarian Sirene cheese traditionally used in this recipe is replaced by white Cheshire, Lancashire or white Stilton.

- ***Preparation: 20 minutes***
- ***Cooking: 1 1/2 hours***

150g/5oz coarse yellow cornmeal
1 1/2tsp salt
50g/2oz butter
225g/8oz white Cheshire, Lancashire or white Stilton, crumbled, or 225g/8oz streaky bacon, grilled and chopped
tomato sauce (optional)
For grilling:
50g/2oz mature Cheddar cheese, grated

- ***Serves 6*** 
- ***340cals/1430kjs per serving***

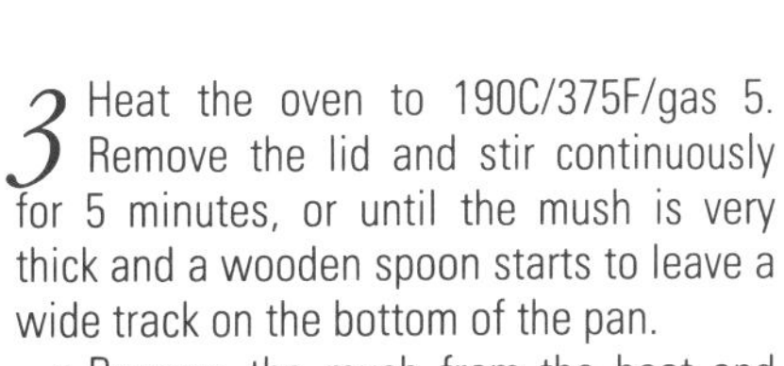

1 Put the cornmeal in a large frying-pan and dry-fry it over moderate heat, stirring all the time, for 3-4 minutes, or until it looses its yellowness and acquires a very light-beige colour.

2 Pour 1L/1¾pt water into a saucepan. Pour in the hot cornmeal – on contact with the cold water it should produce a hissing sound – and add the salt. Simmer for about 5 minutes over moderate heat, stirring occasionally, until the mush begins to sputter. Cover the pan, reduce the heat and let it cook for 20 minutes, stirring occasionally.

3 Heat the oven to 190C/375F/gas 5. Remove the lid and stir continuously for 5 minutes, or until the mush is very thick and a wooden spoon starts to leave a wide track on the bottom of the pan.

4 Remove the mush from the heat and add the butter, stirring as it melts. Lightly and quickly stir in the cheese or bacon and season to taste. Butter a round 20cm/8in baking dish and turn the mixture into the dish. Bake on the top shelf of the oven for 40 minutes or until a pale skin forms on the surface. Leave to cool and set for about 1 hour.

5 To serve hot as a main-course dish heat the grill. Slice the baked cornmeal into wedges like a cake, sprinkle with the grated cheese and grill until the crust is crisp and golden brown and the cornmeal is heated through. Serve at once on hot plates, with a simple tomato sauce if you wish.

Cook's tips

For a cold side dish to accompany sliced cold meat, cut the baked cornmeal into squares or oblong shapes and serve with butter and a cabbage or lettuce salad.

PLUMS ALL ROUND

Luscious deep purple plums abound in the Balkans. They form the basis for that powerful head-swimmer known as *ţuică* in Romania, *šlivovica* in Yugoslavia and *slivova rakiya* in Bulgaria. Drink with caution – or, safer still, restrict yourself to homemade plum preserve which is thicker and fruitier than any shop-bought jam.

Apple nectar

A luscious drink for a warm day, serve it on its own, with biscuits or a slice of cake. An ideal way to cope with a glut of cooking apples.

- ***Preparation: 20 minutes***
- ***Cooking: 1¼ hours***

1kg/2¼lb quartered, peeled and cored cooking apples (about 1.4kg/3lb)
400g/14oz sugar
3 cloves
juice of 1 medium-sized lemon, strained

- ***Makes 1.8L/3pt nectar*** 
- ***325cals/1365kjs per 300ml/½pt***

1 Heat the oven to 110C/225F/gas ¼. Choose enough narrow mouthed glass bottles with plastic-lined lids to hold 1.8L/just over 3pt liquid. Put the bottles and lids, a soup ladle and a plate on a baking tray in the oven for 30 minutes or until ready to fill.

2 Meanwhile put a large heat-resistant plastic kitchen funnel on its side in a ▶

large pan and add enough water to immerse it. Cover the pan, bring the water to a rolling boil and boil for 10 minutes. Remove the pan from the heat and reserve the funnel in the water.

3 Pour 300ml/10fl oz water into an enamelled or stainless steel pan and add the apples and the cloves. Cover and simmer over low heat for 30 minutes, stirring occasionally. Discard the cloves. Purée the apple pulp in a blender or push it through a nylon sieve. Return the purée to the enamelled or stainless steel pan.

4 In a large pan bring the sugar and 1L/1¾pt water to the boil over low heat, stirring until the sugar has dissolved, then boil for 2 minutes. Pour the clear syrup, together with the lemon juice, into the purée, mixing thoroughly. Return the pan to the heat and cook gently for 10 minutes, stirring occasionally.

5 Adjust the heat under the purée to a bubbling simmer. Using protective gloves, transfer the hot plate from the oven to the top of the cooker close to the pan. Put the bottles, one at a time, on the plate. Shake the water off the funnel and set it in a bottle neck, then ladle the bubbling nectar through the funnel, filling the bottle up to the top. Immediately screw down the cap and leave to cool. Repeat with the remaining bottles.

6 Serve well chilled, shaking the bottle before pouring the nectar into small glasses. The nectar will keep in a cool dark place for about 4 months.

Cook's tips

Collect bottles with pretty shapes and fill them with nectar to make unusual Christmas presents.

Maraschino ice

If maraschino liqueur is not available, use cherry brandy or kirsch.

- ***Preparation: 8 minutes, plus 4 hours freezing***

600ml/1pt thick cream, well chilled
200g/7oz icing sugar, sieved
4tbls maraschino liqueur
For the garnish:
maraschino liqueur
6 cocktail maraschino cherries
a piece of angelica, cut into 'leaves'

- ***Serves 6*** 
- ***570cals/2395kjs per serving***

1 If using the freezing compartment of a refrigerator turn it to its highest setting about 2 hours before you start.

2 Whisk the cream, sugar and liqueur together until softly stiff. Pour the mixture into an 850ml/1½pt container, cover and freeze for about 4 hours.

3 Serve the ice cream straight from the freezer, scooping it into glasses, with a little liqueur sprinkled on. Decorate each portion with a maraschino cherry and angelica 'leaves'.

Menu Milanese

Some of Italy's best – and most well known – regional dishes come from Lombardy. Milan, the original capital of the region and also its gastronomic centre, offers specialities such as minestrone and ossobuco.

Italian cooking is surprisingly varied; each area has its own specialities, based on the ingredients produced locally. In the cooler, northern regions a cuisine has developed that is more substantial and sophisticated than the straightforward styles of cooking based on vegetables and roasts found further south. As well as climate, foreign influences have also left their mark on Italy's cuisine.

The Saracens introduced rice into Italy in the 8th century. In the southern part of Lombardy, which stretches from the Alps to the River Po valley, rice, rather than pasta, still provides the staple diet.

In the 19th century, first the French and then the Austrians invaded Lombardy. The French contributed a great deal to Lombard cooking and the adoption of French methods was made easier by the fact that Lombard cooking, like most French cooking, was already butter- rather than oil-based. But the great lesson taught by the French was that food should be both carefully and precisely prepared.

Slow cooking

The Austrians left a preference for slow, carefully cooked pot roasts and for braising, stewing and boiling meats. One of the enduring rules of the Milanese kitchen is that food is cooked with as little water as possible and at a very low temperature for a long time. This prevents meat and vegetables from drying up and produces tender food cooked in its own natural juices.

Gorgonzola pasta (page 103)

A touch of the blues

Milk is used in much of Lombard cooking and a great deal of cheese is made, providing one of the main sources of income for many farmers. Gorgonzola cheese takes its name from a small town in northern Lombardy which, historically, was

invaded every year by herds of cattle coming down from the summer pastures to winter in the southern plains. When the herds stopped in Gorgonzola to rest, the local people began to produce cheese to use up the milk.

Other famous Lombard cheeses include Bel Paese and Grana Lodigiano, a slightly inferior version of Parmesan cheese. Grana Lodigiano comes from the town of Lodi where more cheese is made than in Parma.

The woods in Lombardy yield chestnuts, mushrooms and game, and the streams are full of trout, carp, perch, pike, eels and many other kinds of fish. Grapes for wine are grown on the foothills of the Alps. In the fertile valley of the River Po, excellent fruit, vegetables, sugar beet and cereals are produced.

Rice fields extend from Milan to the city of Turin, so it is not surprising that in Milan's famous Minestrone rice replaces the pasta used in similar vegetable soups elsewhere in Italy. As you go further north in Lombardy, the rice fields give way to maize. Here polenta, which is prepared by stirring ground maize into boiling water, is common.

In a shop window, an array of traditional breads of the region

Minestrone

- ***Preparation: 50 minutes***
- ***Cooking: 3½ hours***

50g/2oz butter
50g/2oz pancetta or unsmoked streaky bacon, chopped
3 onions, sliced
4 carrots, diced
2 celery stalks, diced
350g/12oz potatoes, diced
150g/5oz dried borlotti beans, soaked overnight, or 400g/14oz canned cannellini beans
2 courgettes, diced
100g/4oz French beans, diced
100g/4oz shelled peas
200g/7oz Savoy or other green, leafy cabbage, shredded
1.5-1.7L/2½-3pt meat stock or 3 stock cubes dissolved in the same quantity of water
200g/7oz tomatoes, skinned, or canned plum tomatoes
salt and pepper
175g/6oz medium-grain rice, preferably Italian
40g/1½oz freshly grated Parmesan cheese

- ***Serves 6*** 
- ***410cals/1720kjs per serving***

1 Melt the butter in a saucepan large enough to hold all the ingredients. Add the pancetta or bacon and the onion, and sauté them over low heat until soft.

2 Add the carrots and celery to the pan and, after 2-3 minutes, stir in the potatoes and drained dried beans, if using. After 2-3 minutes, stir in the courgettes, French beans and the peas. After 5 minutes, stir in the cabbage. Cook for 5 minutes, stirring.

3 Add the stock, tomatoes and salt and pepper to taste. Cover and cook over very low heat for about 3 hours.

4 Add the rice, stir, and then add the canned beans (if you are using them instead of dried ones). Cook for 12-15 minutes or until tender.

5 Serve the soup piping hot, sprinkled with the Parmesan.

Cook's tips

Minestrone is even better when it is made the day before. In the summer in Milan, it is often served at room temperature.

Spinach cake

- **Preparation: 35 minutes, plus soaking**
- **Cooking: 1 hour**

50g/2oz stale white bread, crusts removed
500ml/18fl oz milk
100g/4oz butter
40g/1½oz flour
450g/1lb frozen spinach, thawed, or 1kg/2¼lb fresh spinach, cooked and cooled
3 eggs
25g/1oz almonds, skinned and chopped
25g/1oz pine nuts, chopped
25g/1oz digestive biscuits, crumbled
25g/1oz sultanas
½tsp ground cinnamon
½tsp fennel seeds
2tbls freshly grated Parmesan cheese
salt and pepper
oil, for greasing
parsley and tomato slices, to garnish

- **Serves 6-8**
- **385cals/1615kjs per serving**

1 Soak the bread in the milk for 5-10 minutes, then break it up with a fork. Heat the oven to 200C/400F/gas 6.

2 Melt 75g/3oz of the butter in a saucepan over low heat, add the flour and cook for 2 minutes, stirring. Add the bread and milk and cook the mixture for 5 minutes, stirring.

3 Squeeze all the water from the spinach. Add the spinach to the saucepan and stir it for 1 minute. Remove the pan from the heat, pour the mixture into a bowl and cool.

4 Beat the eggs lightly and stir them into the spinach mixture. Add all the other ingredients, except the oil and garnishes, then taste and add salt and pepper to taste.

5 Grease a loose-bottomed 24cm/9½in cake tin well with oil. Spoon the mixture into it, dot over the remaining butter and bake for 40-50 minutes or until the cake is brown on top and firm in the centre.

6 When cool, remove the cake from the tin, garnish with parsley and tomato and serve at room temperature.

FASHIONABLY FRENCH

The French invaded Lombardy in the 19th century and contributed greatly to the cooking of the region. So strong was their influence on Lombardy's cuisine that at one stage it became much more fashionable to eat French food in Milan rather than local Italian specialities.

Variations

This recipe is usually served as first course. However, served with a lightly tossed green salad it is ideal as a light lunch or supper dish. For a sweet spinach cake, beat 50g/2oz sugar into the eggs and omit the salt, pepper and cheese.

Italian braised beef

This dish goes well with sliced carrots flavoured with garlic and rosemary

- **Preparation: 30 minutes**
- **Cooking: 2 hours 50 minutes**

65g/2½oz butter
4 large shallots or 2 small onions, chopped
2tbls flour
salt and pepper
1.1kg/2½lb chuck steak or top rump of beef, in 1 piece
3 carrots, chopped
1 celery stalk, chopped
75ml/3fl oz white wine vinegar
350ml/12fl oz milk

- **Serves 6**
- **470cals/1975kjs per serving**

1 Heat the butter in a heavy flameproof casserole over low heat and when hot add the shallots or onions and cook for 5-6 minutes or until soft and pale golden in colour.

2 Season the flour with salt and pepper and then lightly coat the piece of beef with it.

3 Add the meat to the casserole, raise the heat to medium and brown it on all sides.

4 Add the carrots and celery and gently fry them for 5 minutes. Pour in the vinegar and boil briskly until it evaporates; add the milk, cover tightly and simmer for 2-2½ hours.

5 Remove the meat and reserve it. Pureé the vegetables and cooking liquid in a blender or food processor to make a sauce. Taste and adjust the seasoning.

6 Cut the meat into 1cm/½in slices and lay them on a warmed platter. Cover the meat with some of the sauce and serve the rest handed separately in a warmed sauceboat.

Risotto alla milanese

- ***Preparation: 20 minutes***
- ***Cooking: 1 hour***

100g/4oz butter
2 large shallots, or 1 onion, finely chopped
50g/2oz pancetta or unsmoked streaky bacon, finely chopped
450g/1lb medium-grain Italian rice
100ml/3½fl oz dry white wine
1.5L/2½pt hot home-made stock or 2 chicken stock cubes dissolved in the same quantity of boiling water
pinch of saffron strands
salt and pepper
50g/2oz freshly grated Parmesan cheese

- ***Serves 6-8*** 
- ***485cals/2035kjs per serving***

1 Heat 75g/3oz butter in a large saucepan over low heat and gently fry the shallots and the pancetta for 5-6 minutes until the shallots are translucent.

2 Add the rice and cook over medium heat for 3 minutes, stirring continuously with a wooden spoon. Add the wine and cook for 1 minute, stirring, then add 200ml/7fl oz of the hot stock and stir well. When the stock has been absorbed (after 3-4 minutes) add another 200ml/7fl oz stock and continue cooking for 15 minutes, stirring the mixture occasionally.

3 Stir the saffron into the rice mixture, taste and season with salt and pepper, if necessary.

4 Continue adding about 150ml/¼pt hot stock at a time until it is absorbed. (The stock takes 3-4 minutes to be absorbed each time 150ml/¼pt is added.) When all the stock is used, the rice should be creamy but not mushy.

5 When the grains of rice are tender but still firm when tested, add the remaining butter and sprinkle over half the Parmesan cheese. Cover the pan for 2-3 minutes. Mix well and serve with the remaining Parmesan cheese sprinkled over the top.

SPANISH INFLUENCE

Spanish rule in the 17th century resulted in a rice dish which resembled paella more than the famous Risotto alla milanese which is known today.

Above, Italian braised beef (page 101) and right, Ossobuco (page 103) with Risotto alla milanese (see left).

Variations

For rice that is a bright yellow in colour, substitute a large pinch of turmeric for the saffron.

Ossobuco

- ***Preparation: 20 minutes***
- ***Cooking: 2 hours 20 minutes***

65g/2½oz butter
1tbls vegetable oil
1 onion, very finely chopped
6-7 pieces shin of veal, 5cm/2in thick, tied securely together with string (about 1.5-1.6kg/3¼-3½lb)
2tbls flour
salt and pepper
150ml/¼pt dry white wine
about 250ml/9fl oz beef stock
Risotto alla milanese (page 102), to serve

For the garnish:
1tsp grated lemon zest
2 garlic cloves, chopped
2tbls chopped parsley

- ***Serves 6*** 
- ***340cals/1430kjs per serving***

THE SHORT AND MARROW

Ossobuco means marrow bone. The dish is made with pieces of veal shin bone tied together to keep the marrow in place while they cook. The hind shin is meatier and more tender than the front shin and ideally the meat should be from an animal not more than three months old.

1 Heat the butter and oil over low heat in a large heavy frying pan. When hot, add the onion and cook until it becomes translucent in colour.

2 Dredge the bundle of veal in the flour, shaking off any excess. Brown the meat well on all sides in the pan and season with salt and pepper.

3 Add the wine and boil briskly for 5 minutes, turning the bundle over and over during cooking.

4 Add 175ml/6fl oz of the stock, cover the pan tightly and cook over very low heat for about 2 hours. Turn and baste the bundle every 15 minutes. If the sauce becomes dry, add more of the hot stock or water during the cooking. If the sauce is too thin when the veal is done, remove the meat and boil the sauce briskly over high heat to reduce and thicken its consistency.

5 Cut the string and arrange the pieces of veal on a warmed serving dish; pour over the sauce. Sprinkle the lemon zest, garlic and parsley over the meat, mix well and serve piping hot together with Risotto alla milanese.

Pasta shells with Gorgonzola sauce

- ***Preparation: 20 minutes***
- ***Cooking: 15 minutes***

2tbls shelled pistachios
350g/12oz conchiglie
salt and pepper
100g/4oz butter
90g/3½oz Gorgonzola cheese
125ml/4fl oz single cream
2tbls brandy

- ***Serves 6*** 
- ***470cals/1975kjs per serving***

1 Put the pistachios in a bowl and cover them with boiling water. Let them stand for 2 minutes, then drain and peel them. Pound the nuts in a pestle and mortar or grind them in a blender or food processor and reserve.

2 Drop the pasta shells into rapidly boiling, salted water. Stir them with a wooden spoon, bring the water back to the boil and adjust the heat so that the water boils fast without boiling over.

3 Meanwhile, melt the butter and the cheese in a small, heavy saucepan over very low heat. As soon as they are melted, add the cream and cook very gently for 5 minutes, stirring continuously.

4 Remove the saucepan from the heat and add the pistachios and the brandy. Taste the sauce and adjust the seasoning, if necessary.

5 When the pasta shells are tender but still firm, drain and pour the sauce over them. Mix well and serve at once.

Cook's tips

Conchiglie are small pasta shells which look very attractive in pasta dishes. You can buy conchiglie fresh or dried and white, green or pink in colour.

FLOWER POWER

For a pretty decoration for cakes and puddings, brush clean, dry edible flower petals with lightly beaten egg white (lilacs and violets were used here). Sprinkle with caster sugar and leave to dry for a few minutes. You can add fresh green leaves such as ivy, but these must not be eaten!

Italian-style chocolate cake

Italian-style chocolate cake

Traditionally, this cake is served with tiny cups of black coffee

- ***Preparation: 45 minutes, plus chilling***

100g/4oz unsalted butter, softened
100g/4oz icing sugar
50g/2oz cocoa powder, sifted, plus extra for dusting
1 egg
75g/3oz shortbread biscuits
75g/3oz almonds, skinned and chopped
1½tbls brandy
oil, for greasing
150ml/¼pt double cream, whipped

- ***Serves 6*** 
- ***470cals/1975kjs per serving***

1 Beat the butter and the sugar together until creamy. Add the sifted cocoa and mix it in well.

2 Beat the egg and beat well into the cocoa mixture.

3 Break and crush the biscuits with a rolling pin and add the crumbs to the mixture with the chopped almonds. Mix well. Add the brandy and mix thoroughly.

4 Lightly oil a 19cm/7½in sandwich tin and line it with greaseproof paper. Spoon the biscuit crumb mixture into the tin, then press it down and smooth it with a palette knife. Refrigerate the mixture for at least 3 hours.

5 Turn the cake out on a round or oval dish, cover the top with whipped cream and dust with cocoa powder.

SPECIALITY CAKES

Two cakes which are a speciality of Lombardy are famous world-wide – *colomba* and *panettone*. Colomba, traditionally eaten at Easter and shaped like a dove, is made from a fine, light dough, which is baked and covered with sugar crystals and toasted almonds. Panettone is a large, dome-shaped cake which finds its way into most Italian households at Christmas. Delicious with coffee or sparkling wine, it is made from a similar dough to colomba and includes lots of sultanas and candied citrus peel.

The Marvellous Midi

Throughout France, plentiful produce provides a base for its cuisine – nowhere more so than the sun-baked region of Languedoc-Roussillon, locally named 'le Midi'

Rustic liver terrine (page 106)

THE COOKING OF FRANCE is wonderfully varied – each region has developed its own style based on the local produce. In the north, Normandy is famous for its cream apples and seafood; while to the east, in Picardi, the cooking is less rich and fairly robust, with hearty soups and gamey pâtés. Further south, in Savoy, the cooking is simpler, though it still relies on fresh ingredients; while Bordeaux and Alsace, like other great wine producing areas, are renowned for dishes which use wine in cooking.

To the south-west, in the huge region of Languedoc-Roussillon, which borders the Mediterranean and extends from the Pyrenees to the river Rhône, the climate is largely responsible for the richness of produce to be found all over the area: wines from the shapely hills of Minervois, the valleys of the Aude and the lower Rhône; beautiful early vegetables like asparagus and artichokes; a particularly flavoursome garlic; many varieties of wild mushrooms and herbs; and that rare and highly prized delicacy – the aromatic truffle. An abundance of delicious fruit grows here, from plums, apricots and grapes to strawberries, oranges and figs.

Hearty fare

Livestock is equally plentiful, from the mountain lamb and mutton to the veal and beef of the pasture lands and the backyard pig. Game there is in plenty. Domestic poultry is a succulent speciality – fat capons, geese, ducks and turkeys all feature in local recipes and their carefully fattened livers *(foie gras)* are reserved for making pâtés and terrines.

As well as the numerous types of fish from the Mediterranean and shellfish cultivated in the salt lakes, the Languedoc rejoices in a good variety of freshwater fish that are native to its many streams and rivers: tench, bream, perch, pike and shad as well as river trout.

The main dishes most widely enjoyed tend to be substantial and robust despite the heat. Soups like *pot au feu carcassonais* (with ribs of beef) and the fish soups of Sète and Collioure are almost meals in themselves. Meaty fish like tuna and cod are prepared in rich ragoûts, or stuffed; mutton and beef are braised for hours.

Garlic for breakfast

Once part of the independent province of Catalonia, Roussillon still maintains many of its ancient Catalan traditions and has many culinary specialities of its own.

Most famous of these is the substantial thick cabbage, pork, haricot bean and game soup, *braou bouffat* ('good eating'). Partridge with morels is another speciality. Cerdagne is one of the few regions where another rare and delicious dish, *civet d'isard* (mountain goat) may be found.

Garlic and olive oil are much beloved by the Roussillon people and are eaten even first thing in the morning. A typical breakfast will be slices of country bread rubbed with garlic and moistened with olive oil – and that's just the start of the day!

Rustic liver terrine

- ***Preparation: 1½ hours, plus cooling and chilling***
- ***Cooking: 3½ hours***

450g/1lb calf's liver
100g/4oz lean pork
100g/4oz salt pork
175g/6oz chilled lard
175g/6oz fresh white breadcrumbs
100ml/3½fl oz dry vermouth
50g/2oz chopped parsley
¼-½tsp ground mace
¼-½tsp ground cloves
salt and pepper
450g/1lb goose, turkey or chicken livers
1 bay leaf
3 or 4 mushrooms, sliced
90g/3½oz flour
hot toast or French bread, to serve

- ***Serves 8*** 
- ***620cals/2605kjs per serving***

1 Chop the calf's liver, lean and salt pork and 100g/4oz lard fairly finely (do not mince – the texture should be rough). Chop the remaining lard and reserve.

2 Soak the breadcrumbs in the vermouth and add the mixture to the chopped ingredients. Mix well, adding the parsley, mace, cloves and salt and pepper to taste and reserve. Heat the oven to 150C/300F/gas 2.

3 Coarsely chop the goose, turkey or chicken livers. Pack a layer of the calf's liver mixture into the bottom of a round or oval 1.2L/2¼pt lidded terrine. Cover the calf's liver with a layer of goose, turkey or chicken livers. Repeat the layering process, pressing each layer down; end with a layer of mixed ingredients.

4 Place a bay leaf in the centre and surround with the mushroom slices. Top with a thin covering of the rest of the lard. Mix the flour and 4tbls water to form a smooth, lump-free paste and seal on the lid with this paste.

5 Place in a pan of hot water to come halfway up the side of the terrine and cook in the oven for 3½ hours.

6 Remove from the oven, remove the lid, cool quickly, then cover lightly and chill thoroughly. Unmould and wrap the terrine in a 'collar' of foil or serve from the dish with slices of hot toast or crusty French bread.

Garlic in abundance – hardly any savoury 'Midi' dish can do without it

Braised beef with garlic

- ***Preparation: 15 minutes***
- ***Cooking: 3¾ hours***

3tbls olive oil
1 large onion, chopped
1.4kg/3lb topside or top rump, rolled and tied
2tbls seasoned flour
300ml/½pt dry white wine
225g/8oz button mushrooms
2 large tomatoes, coarsely chopped
2 heads of garlic, cloves peeled but left whole
salt and pepper

- ***Serves 6-8*** 
- ***470cals/1975kjs per serving***

1 Heat the oven to 150C/300F/gas 2. Heat the oil in a large flameproof casserole, add the onion and fry for 5 minutes, stirring.

2 Roll the meat in the seasoned flour, then add it to the casserole and brown it on all sides over high heat.

3 Add the wine, stir for 2-3 minutes, then add the mushrooms, chopped tomatoes, garlic cloves, a little salt and lots of pepper. Cover and cook in the oven for 3½ hours or until very tender.

4 Transfer the meat to a heated serving platter, remove the string and carve the joint into thick slices with a sharp knife. Keep them warm.

5 Put the casserole over medium heat and boil the sauce for 4-5 minutes until it is really thick. Spoon some of the sauce over the slices of meat and serve the rest separately in a sauceboat.

Trout with cream and mushrooms

- ***Preparation: 20 minutes***
- ***Cooking: 25 minutes***

4 trout, 225g/8oz each, cleaned
salt and pepper
seasoned flour
100g/4oz butter
250g/9oz mushrooms, thinly sliced
4tsp lemon juice
300ml/½pt double cream
pinch of dried chervil

- ***Serves 4***
- ***675cals/2835kjs per serving***

1 Lightly salt the trout inside and coat them in seasoned flour.

2 Melt 40g/1½oz butter in a saucepan over a low heat, tip in the mushrooms and turn them until slightly softened, about 3-4 minutes. Add the lemon juice, season and reserve.

3 Warm a flameproof serving dish into which the trout will fit comfortably side by side. Gently melt the rest of the butter in a large frying pan and fry the trout for 4 minutes on each side. Transfer the trout to the serving dish and add the mushrooms and their juices to the frying pan, pour over the cream, sprinkle on the chervil and raise the heat slightly.

4 Meanwhile, heat the grill to medium-high. When the cream is bubbling, remove the frying pan from the heat and pour the sauce over the trout. Put the serving dish under the grill. Continue cooking for about 10 minutes, occasionally basting the exposed sides of the trout to prevent burning. Serve at once.

Serving ideas

Serve the trout with fluffy mashed potatoes, plain boiled new potatoes or rice, accompanied by a crisp green salad.

CONFIT TERRITORY

The *confit* is a French culinary term for one of the oldest forms of preserving food and is a speciality of south-western France. A piece of pork, goose, duck or turkey is cooked in its own fat, stored in a pot and covered in the same fat to preserve it.

CASSOULET CASEBOOK

There are three main regional versions of cassoulet – from Castelnaudary, Carcassonne and Toulouse. This recipe is based on the Castelnaudary version which seems to have the best claim as originator. The Carcassonne variation adds lamb or mutton, sometimes even partridge; Toulouse adds mutton or lamb, too, plus its famed chunky sausage and goose confit.

Traditional cassoulet

- ***Preparation: 1 hour, plus overnight soaking***
- ***Cooking: 3½ hours***

100g/4oz lard
800g/1¾lb duck or goose joints
6 onions, chopped
6 garlic cloves, crushed
100g/4oz tomato purée
400g/14oz canned tomatoes
2tsp paprika
pinch of cayenne pepper
salt and pepper
2 bouquets garnis
500g/18oz lean pork shoulder, cut into 2.5cm/1in cubes
6 shallots, chopped
25g/1oz flour
300ml/½pt beef stock
stick of cinnamon
800g/1¾lb haricot beans, soaked overnight
1 large onion, stuck with 6 cloves
1 large carrot
2 whole heads of garlic
1 pork or beef shin bone (optional)
500g/18oz salt pork belly, in one piece
200g/7oz garlic sausage, in one piece
1 extra garlic clove, peeled
6-8 'country-style' sausages
75g/3oz fresh breadcrumbs

- ***Serves 8-10***
- ***1110cals/4660kjs per serving***

1 Melt the lard in a large, heavy saucepan over medium heat, add the duck or goose joints and fry until golden, 4-5 minutes. Remove the joints and reserve. Pour off three-quarters of the fat and reserve.

2 To the fat remaining in the pan, add the chopped onions and crushed garlic and stir for 2-3 minutes over medium heat. Stir in the tomato purée, add the canned tomatoes, paprika, cayenne, salt and one bouquet garni. Return the duck or goose to the pan, cover and cook over a low heat for 2½-2¾ hours.

3 Meanwhile, in another heavy pan, melt 2tbls of the reserved fat and lightly fry the pork shoulder. When evenly coloured, add the shallots, sprinkle on the flour, stir for 1-2 minutes, and gradually add the stock, still stirring. Season with salt and pepper, add the cinnamon, cover and simmer gently for 2-2½ hours.

4 Meanwhile, drain the beans, put in a large saucepan, cover with fresh cold water, add the onion stuck with cloves, the carrot, garlic heads, bone, if using, salt pork and the second bouquet garni. Bring to the boil and simmer for 1½ hours.

5 Season the beans with salt, add the garlic sausage and simmer for a further 20 minutes.

6 Remove the pans containing the meats from the heat, uncover, check the seasoning (which should be rather strong) and reserve.

7 Remove the salt pork and garlic sausage from the beans and set aside. Drain the beans and discard the onion, carrot, bone, garlic and bouquet garni. Thickly slice the sausage.

8 With a slotted spoon, remove the duck and pork pieces from their respective sauces and set aside. Discard the bouquet garni and the cinnamon stick, combine both sauces with the drained beans and reserve.

9 Rub a large earthenware or cast iron casserole with the remaining garlic clove, then spoon a generous layer of the bean mixture into the dish. Lay two or three pieces of duck or goose on top of the beans, then some of the pork cubes and three or four pieces each of the salt pork and garlic sausage. Cover with another layer of beans and repeat the process, finishing with a layer of beans. Heat the oven to 230C/450F/gas 8.

10 Put the remaining reserved fat into a frying pan and fry the sausages until they are three-quarters done. Tuck the sausages into the top layer of beans and sprinkle with fat from the frying pan and the breadcrumbs.

11 Put the casserole near the top of the oven; cook for 20 minutes or until the crumbs are nicely browned and the whole dish well heated through. Serve immediately with crusty French bread and a crisp green salad.

Partridges with Seville oranges

- **Preparation: 30 minutes**
- **Cooking: 1¾ hours**

4tbls olive oil
1 large Spanish onion, finely chopped
4 young partridges
2tbls flour
150ml/¼pt medium-dry white wine
8 Seville oranges, peeled with all pith removed
juice of 2 Seville oranges
300ml/½pt chicken stock
1 bouquet garni
2 bay leaves
6 garlic cloves, peeled but whole
200g/7oz canned pimientos, drained and sliced
salt and pepper
sprigs of watercress, to garnish

- **Serves 4** 
- **645cals/2710kjs per serving**

1 Heat the oil in a large, heavy saucepan over medium-high heat and sauté the onion for 4-5 minutes. Remove with a slotted spoon and reserve.

Grape tart (page 110)

OF GOOSE AND GANDER

The goose is of great importance in Languedoc, which is one of France's main *foie gras* areas. *Foie gras* is the enlarged liver of a goose that has been force-fed up to the point at which further cramming has no effect. Because geese are a stubborn species and refuse to breed more than once a year, the season for fresh fattened goose liver only starts in early winter, its peak coinciding with the increased Christmas demand.

2 Add the partridges and brown on all sides. Stir in the flour and cook for 2-3 minutes. Pour in the wine, let it bubble for 2 minutes, then stir.

3 Add the oranges, orange juice, chicken stock, bouquet garni, bay leaves, garlic cloves and reserved onions. Cover with a tightly fitting lid, turn the heat down to very low and simmer gently for 1 hour.

4 Stir in the pimientos, cover and cook for a further 20-30 minutes – the oranges should be very soft and have absorbed as much of the sauce as they can. Season to taste. Transfer to a warm serving dish, garnish and serve.

Cook's tips

Although Seville oranges appear only briefly for a few weeks in the winter, they freeze well. Just rinse and dry the whole fruits and freeze in a freezer bag.

Variations

You can use small sweet oranges instead of Seville oranges – but use one sweet orange and one lemon for the juice to match the taste of the original dish.

Grape tart

- *Preparation: making and resting the pastry, then 30 minutes*
- *Cooking: 45 minutes*

Rich shortcrust pastry (see Cook's tips)
350g/12oz large green grapes
flour, for sprinkling
a little egg white, for brushing
50g/2oz caster sugar
3tbls greengage, quince or crab-apple jelly

- *Serves 4-6* 
- *650cals/2730kjs per serving*

1 Make the rich shortcrust pastry and set aside in a cool place to rest for 15 minutes. Heat the oven to 190C/375F/gas 5.

2 Remove the seeds from the grapes by slicing them three-quarters through from the stalk end and easing out the seeds with the point of a knife. Stand the grapes, cut side down, on a clean cloth.

3 Roll out the pastry on a lightly floured surface and line a 20cm/8in flan tin with it. Prick the bottom with a fork, brush with the egg white and fill with the grapes, pressing them close together, cut sides down. Sprinkle with the sugar and bake for 35-45 minutes.

4 As soon as the tart is baked, heat the jelly until liquid and brush over the tart. Serve warm or cold.

Apricot and rice meringue

- *Preparation: 15 minutes*
- *Cooking: 1 hour 20 minutes*

450g/1lb ripe apricots, halved and stones removed
100g/4oz short-grain rice
600ml/1pt milk
1 vanilla pod
2 strips of lemon zest
225g/8oz caster sugar
2 eggs, separated

- *Serves 4-6* 
- *480cals/2015kjs per serving*

1 Cover the apricots with water, bring to the boil and simmer until very tender, about 10 minutes. Drain and reserve.

2 Wash the rice thoroughly in a sieve under hot running water.

3 Bring the milk to the boil in a large saucepan with the vanilla pod and the lemon zest, add the rice, reduce the heat, cover and simmer very gently for 25 minutes. Add half of the sugar, cover and simmer for 25 minutes, stirring occasionally to prevent the rice from sticking. Allow the rice to cool slightly, then beat in the egg yolks.

4 Turn the rice into an 850ml/1½pt shallow ovenproof serving dish. Heat the oven to 180C/350F/gas 4. Put the apricots in a layer on top of the rice.

5 Whisk the egg whites in a large, clean, dry bowl until stiff peaks form. Carefully fold in the remaining sugar with a metal spoon, then spoon the meringue mixture into a piping bag fitted with a 1.5cm/½in star nozzle. Pipe the meringue in a lattice pattern over the apricots and bake for 15 minutes or until golden. Serve hot, warm or cold.

Cook's tips

For rich shortcrust pastry, sift together 225g/8oz flour and a pinch of salt into a bowl. Add 150g/5oz diced butter and rub in until the mixture resembles breadcrumbs. Stir in 1tbls cold water and stir in to make a soft dough. Add 1-2tsp more cold water, if necessary. Knead lightly and pat into shape.

ALMONDS FROM ARABIA

The use of almonds in various regional dishes is part of a tradition that goes back to the Arabs who spilled over into the Languedoc from neighbouring Spain centuries ago.

Pounded almonds are an ingredient of the sauce that accompanies snails in *escargots à la narbonnaise,* and almonds lend a special flavour to the local *bouillabaisse* (fish soup).

Almonds also figure largely in the filling for the traditional sweet flan known as *croustade languedocienne* and the huge repertoire of sweetmeats and pastries of which the Languedoc is so proud. Specialities include honey croquettes from Narbonne, Limoux cakes and the chocolates and *marrons glacés* of Montpellier.

Flavours of Northern India

From elaborate meat and poultry dishes to simpler vegetarian fare, the cooking of northern India is full of flavour and often highly spiced

Tandoori chicken (page 112)

THE TRADITIONS AND taboos of two great religions shape the cooking of northern India. Both Muslims and Hindus have strict rules about the types of food eaten as well as preparation and serving methods. As a result, the area offers two very different, but equally interesting cooking styles.

The Muslims of northern India eat the same kind of food as those in neighbouring Pakistan; this includes meat and many elaborate and highly spiced dishes. Hindu cooking, on the other hand, is much simpler than that of the Muslims and is basically vegetarian. Occasionally meat is eaten, but never by women, and no Hindu, man or woman, will ever touch beef.

Before the Muslim conquest, when the Hindus ruled India, large quantities of meat were eaten. The Hindu emperor Asoka, who ruled in the 3rd century BC, wrote that before he became a Buddhist and gave up killing animals, thousands of birds and beasts were slaughtered for his kitchen. Even so, among Hindu princes, the eating of meat carried on for centuries – it was the priestly and trading classes who gave up the eating of meat from very early times.

Ingredients guide
Spices are an integral part of Indian cuisine. You can buy most spices ready ground for convenience, but for the best flavour, buy whole ones and grind them yourself, with a pestle and mortar or in a clean coffee grinder. In the photo are some of the most popular spices. Top: cumin seeds, dried chillies, coriander seeds, cinnamon sticks. Bottom: cloves, green cardamom pods, black peppercorns.

The rules about eating fish are not so strict and depend on geographic divisions. The Hindus of northern India do not eat fish, while those in eastern India do.

Northern staples

In eastern and southern India the staple food is rice, but in northern India it is unleavened bread. This is usually made from wheat flour although it is often made from millet flour among the poorer people. Bread, whatever form it takes, is called *roti,* and the expression 'come and eat roti' is a way of asking friends round for a meal.

A Hindu meal in northern India will usually consist of some form of roti such as Paratha (page 113), various kinds of vegetable dishes and *dal* or pulse dishes. These are usually followed by desserts, which are often made from milk, unsweetened milk curds and fruit.

Serving a Hindu meal

In Hindu homes the food is always prepared by the women of the household and is traditionally eaten sitting on the floor on pieces of carpet. Since Hindus believe that contact with food pollutes, the plates and bowls are put on the bare floor, which is then washed after every meal. Food is eaten from a *thali* which can be a silver plate or, in its simplest form, a banana leaf. Each person takes a small amount of food from each of the dishes laid out and places it around the edge of the thali. The centre of the thali is then used to mix the food.

No Hindu touches his or her food with the left hand which is considered unclean. Only the right is used for eating and drinking.

Weighing peppers in Jaipur

Tandoori chicken

In India a whole chicken is roasted in a clay oven or *tandoor.* But as the special flavour comes from the yoghurt marinade, an electric or gas oven does just as well

- ***Preparation: 40 minutes, plus overnight marinating***
- ***Cooking: 1½ hours***

1.6kg/3½lb chicken, cut into portions
3tbls lemon juice
¼tsp chilli powder
salt and pepper
1 large onion, chopped
5cm/2in piece fresh root ginger, peeled and chopped
4-5 garlic cloves, chopped
175ml/6fl oz yoghurt
1tbls ground coriander
1tbls ground cumin
1-2tbls paprika
15g/½oz melted butter

- ***Serves 4***
- ***390cals/1640kjs per serving***

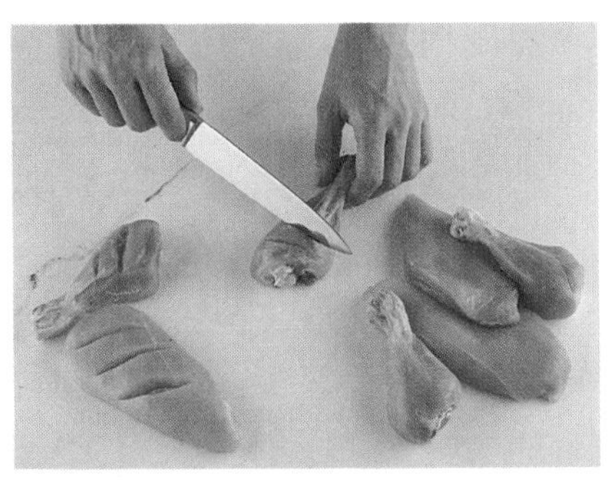

1 Skin the chicken portions, then prick all over with a fork. Slash with a sharp knife. Mix the lemon juice, chilli powder and salt and pepper to taste together in a jug. Pour over the chicken and rub into the slashes.

2 Pound the onion, ginger and garlic to a paste using a pestle and mortar, or blend in a food processor until smooth. Stir in the yoghurt, coriander, cumin and paprika. Brush all the marinade over the chicken, cover and leave to marinate in the refrigerator overnight.

3 Heat the oven to 230C/450F/gas 8. Remove the chicken from the marinade and place in a roasting tin. Mix the melted butter with the marinade remaining in the dish, then use this mixture to coat the chicken again.

4 Cook the chicken for 1¼-1½ hours or until cooked through, basting occasionally with the juices. Serve hot.

Variations

Instead of using a whole chicken, you could use just breast portions or leg portions. If using breast portions, reduce the cooking time to 1¼ hours.

Parathas

This Indian fried bread is made from dough which is rolled and folded several times to create thin layers. Serve the bread hot

- ***Preparation: 35 minutes***
- ***Cooking: 40 minutes***

75g/3oz flour, plus extra for dusting
75g/3oz wholemeal flour
½tsp salt
75g/3oz butter, melted
50ml/2fl oz milk

- ***Makes 4***
- ***275cals/1155kjs per paratha***

1 Sift the flours and salt together. Add 1tbls melted butter and mix well. Mix the milk and 50ml/2fl oz water and stir it into the dough, then divide it into four balls.

2 Flatten a ball of dough with your palm and roll it into a 15cm/6in circle. Brush with a little of the remaining melted butter, then sprinkle over a little flour. Fold the dough into a semi-circle, brush with a little more melted butter and sprinkle with a little more flour.

3 Fold the dough again to join the corners and form a quarter-circle. Pat the dough to make it into a 15cm/6in circle, then repeat the process. Make three more parathas in the same way.

4 Heat a griddle or heavy frying pan over medium heat and brush it with melted butter. When it is hot, fry one paratha, keeping a check on the underside during cooking. As soon as light brown spots appear, spoon a little butter on the griddle around the dough. Press the paratha frequently with the back of a fish slice and turn occasionally until it is golden brown on both sides, about 4-5 minutes each side. Keep warm while you repeat with the rest of the circles.

Variations

You can make parathas using 175g/6oz wholemeal flour, instead of using half plain and half wholemeal.

ON THE SIDELINES

In India meals are often served with a variety of tasty accompaniments. These can range from something simple like cucumber slices seasoned with salt and pepper, cayenne pepper and lemon juice, or a mixture of chopped tomatoes and onions, to pickles which have been left for weeks or months to mature.

Mango chutney is popular in the west, but use sparingly as too much will mask the flavours of the other dishes. At the other end of the scale, lime pickle has much more of a bite to it, being very hot. There are lots of different pickles and chutneys available from supermarkets and specialist shops, so experiment with new flavours.

A cooling dish known as raita is nearly always included, made from yoghurt combined with vegetables or fruit. Plain or spicy poppadums are often served, and breads such as chapatis, parathas and naans are a good foil for spicy foods.

Lamb kheema kebabs

- ***Preparation: 10 minutes, plus 30 minutes marinating***
- ***Cooking: 20 minutes***

1tbls oil
1 small onion, finely chopped
2 garlic cloves, crushed
5cm/2in fresh root ginger, peeled and finely chopped
1-2 green chillies, seeded and finely chopped
1tsp cumin seeds, lightly crushed
450g/1lb lean minced lamb
1 small egg, lightly beaten
salt and pepper
lemon wedges and Fresh mint chutney, to serve

- ***Serves 4*** 
- ***240cals/1010kjs per serving***

1 Heat the oil in a saucepan and fry the onion, garlic, ginger and chillies over medium heat for 3-5 minutes or until the onion is soft, stirring constantly. Add the cumin seeds. Increase the heat slightly and cook, stirring, until they start to pop. Remove the pan from the heat and leave to cool.

2 Put the lamb in a large bowl, add the onion and spice mixture and the beaten egg, season with salt and pepper to taste and mix well. Chill for 30 minutes.

3 Divide the mixture into eight equal portions. With wet hands, mould each portion into an oval shape around 20cm-25cm/8in-10in flat metal skewers, pressing firmly and leaving at least 2.5cm/1in skewer exposed at each end.

4 Heat the grill to high. Place the kebabs on a grill rack and cook for 12-15 minutes or until browned and cooked through, turning once. Transfer the kebabs to a heated serving plate and serve with lemon wedges and Fresh mint chutney.

Cook's tips

Handle the kebabs carefully while cooking, as they are quite soft in texture and are liable to crumble.

Serving ideas

The kebabs can be arranged on a bed of plain boiled rice and served with a crisp green salad to complete the meal.

Spinach bhajee

- ***Preparation: 25 minutes***
- ***Cooking: 20 minutes***

2 onions
1tbls oil
700g/1½lb spinach
1 green chilli, seeded and finely chopped (optional)
salt
2tbls ghee or butter
2.5cm/1in fresh root ginger, peeled and cut into thin strips

- ***Serves 4*** 
- ***85cals/355kjs per serving***

1 Finely chop one of the onions. Heat the oil in a large, heavy saucepan over medium heat and cook the chopped onion for 5-7 minutes or until golden brown, stirring frequently.

2 Add the spinach and chilli, if using, and season to taste with salt. Stir once, cover and cook over a low heat for 10 minutes, or until the spinach is just tender.

3 Meanwhile, thinly slice the remaining onion. Melt the ghee or butter in a small pan and cook the sliced onion and ginger over medium heat for 5-6 minutes or until crisp, stirring frequently.

4 Serve the spinach piping hot, sprinkled with the onion and ginger.

Fresh mint chutney

- ***Preparation: 5 minutes, plus 30 minutes chilling***

a handful of fresh mint
1 small onion, quartered
1 garlic clove
juice of 1 large lemon
salt
mint sprig, to garnish

- ***Serves 4*** 
- ***5cals/20kjs per serving***

1 Put the mint, onion, garlic and lemon juice in a blender. Season with salt to taste and blend to a smooth paste. Pour the mixture into a small serving bowl, cover well and chill for about 30 minutes.

2 Serve the chutney chilled, garnished with a sprig of mint.

Plan ahead

The chutney can be made in advance and will keep for up to 3 days when stored in the refrigerator.

Variations

Add 1tsp chilli powder, 4tbls yoghurt and a twist of pepper to the basic ingredients; blend to a smooth paste. Dissolve 2tsp honey in 2tbls cider vinegar and stir into the yoghurt mixture. Chill for 1 hour.

Lassi

• *Preparation: 10 minutes, plus chilling*

300ml/½pt yoghurt
300ml/½pt milk
2 cardamom pods, seeded
4 drops of rosewater
3tbls caster sugar
ice cubes
1tsp cinnamon, to garnish

• *Serves 4* 

• *125cals/525kjs per serving*

1 Put the yoghurt, milk, cardamom seeds, rosewater, sugar and 600ml/1pt water into a blender or food processor and blend at high speed until all the ingredients are well mixed. Transfer to a jug, cover and chill well.

2 Add ice cubes and stir well before serving, sprinkling each drink with a pinch of cinnamon.

Cook's tips

Lassi is widely drunk throughout India, either salted as a thirst quencher, or sweetened as a special treat. It is ideal with spicy food.

Carrot halva

• *Preparation: 25 minutes*

• *Cooking: 1 hour*

300g/11oz carrots, grated
5 whole cardamoms
450ml/16fl oz milk
3½tbls ghee or oil
3½tbls caster sugar
2tsp chopped pistachio nuts
2tsp chopped almonds

• *Serves 4* 

• *255cals/1070kjs per serving*

1 Put the carrot, cardamoms and milk in a large, heavy saucepan. Bring to the boil. Lower the heat slightly and continue to cook at a rolling boil, stirring occasionally, until all the milk has evaporated.

2 Heat the ghee or oil in a separate saucepan. Transfer the carrot and milk mixture to it and fry, stirring, for 10 minutes. Add the sugar and stir until dissolved, then serve hot or cold, garnished with the chopped nuts.

Plan ahead

The halva can be made several days in advance and kept in the fridge. Allow it to return to room temperature before serving.

CUCUMBER COOLER

Its cooling properties make raita an excellent 'fire extinguisher' to counteract very hot and spicy food. It can be served with almost any Indian meal. To make a cucumber raita, beat together 300ml/½pt yoghurt, ¼tsp salt and ¼tsp ground cumin. Combine with a 5cm/2in piece of cucumber, cut into strips. Pour into a serving bowl and chill for 2 hours before serving, sprinkled with ¼tsp ground cumin.

Exotic Taste of India

The contrast of fiery dishes with delicately spiced and perfumed fare makes the cooking of Bengal and central India particularly intriguing to western palates

THE EXOTIC TASTE of Central India and Bengal – India is so vast and varied a country that no one particular style of cooking can be regarded as typical. The food which is identified with India, in the west, is really Muslim-style cooking. Because of their cultural heritage Muslim cuisine is based on a mixture of Arab, Turkish and Persian cooking. Europeans, in India, had only Muslim cooks, who prepared Muslim dishes unless they were taught Western cooking. Consequently the Europeans considered that Muslim cuisine was the native cooking of India. The most popular Indian dish in the west is 'curry' which is not Muslim in origin but an Anglo-Indian creation. It is a Tamil word meaning 'a mixture of spices' and is correctly used in the phrase 'curried mutton'; mutton cooked with a spice mixture. Truly Indian cooking varies enormously from region to region and is predominantly Hindu. There is a remarkable difference in culinary style between eastern India and the rest of the country. In the east and more especially in Bengal, the Hindus, including those from the highest castes, eat both fish and meat, whereas most other regions are strict vegetarians. However even the meat-eating Bengalis do not eat beef or pork; Hindus regard the cow as sacred and pork is considered unclean by Muslims and Hindus.

The cooking traditions in the

homes of the Bengali upper classes, who have always cared for good living, serve as a model for the rest of eastern India. Quantity as well as quality is important and the wide range of dishes are served in a similar order to the European way: a tasty appetizer, followed by vegetables, pulses, fish and meat. Although rice or bread is considered the main item, the cooked dishes are the accompaniments. These courses are followed by sweet desserts made with yoghurt or milk.

Vegetarian India

In central and indeed the rest of India the cuisine is predominantly vegetarian, based on the availability of local crops, of which there is an enormous variety, particularly in the central states. These crops consist of wheat, gram and various other pulses, rice, vegetables such as aubergines, garlic, sweet potatoes, cucumbers, okra, squashes and tropical fruits. Indian vegetarian cuisine is interesting and varied, pulses and vegetables are either cooked simply or with delicious blends of spices. Desserts are often made with a mixture of flour and curd cheese, or khoa (whole milk, simmered to a thick sweet mass).

The exception to the eating habits of central India are the Muslims who live in the state of Uttar Pradesh. Their palates have been strongly influenced by the opulent Moghul cuisine, demonstrated by dishes such as their rich meat birianis.

An Indian flower market showing the making of garlands

Fish with mustard

- *Preparation: 5 minutes*
- *Cooking: 15 minutes*

700g/1½lb fish steaks or 3 small trout
salt
2tsp ground turmeric powder
125ml/4fl oz vegetable oil
1tsp panch foran (see Cook's tips)
1 onion, sliced lengthways
1tbls black mustard seeds, ground and mixed with water to make a paste
1tbls strong mustard
¼tsp chilli powder (optional)
2 green chillies, slightly slit

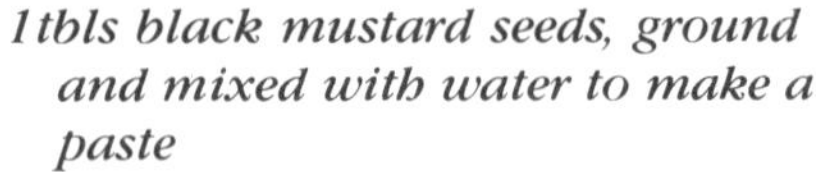

- *Serves 4* 
- *410cals/1720kjs per serving*

1 Wash and dry the fish steaks or trout and sprinkle with salt and ½tsp turmeric powder. Heat half the oil in a frying pan over medium heat. Fry the fish lightly on both sides for about 5 minutes, but do not let it brown. Remove from the pan.

2 Add the rest of the oil to the frying pan and when hot, add the panch foran, sliced onion, mustard paste, the rest of the turmeric, salt to taste, the sugar, chilli powder if using and chillies. Stir these for 30 seconds, add 175ml/6fl oz water and the fish and continue cooking until the liquid is thick. Remove the chillies. Serve with boiled rice.

Cook's tips

Panch foran is a combination of various seeds. Mix together equal quantities of onion seeds, anise seeds, fenugreek and black or white mustard seeds.

Spicy tomato pulse soup

- *Preparation: 30 minutes, plus soaking*
- *Cooking: 40 minutes*

100g/4oz arhar dal
50g/2oz mong dal
2tbls butter
¾tsp cumin seeds
½tsp mustard seeds
1 onion, finely chopped
15mm/½in piece of fresh root ginger, finely chopped
1 garlic clove, crushed
1 green chilli, finely chopped (optional)
2 medium-sized tomatoes, finely chopped
1¼tsp salt
½tsp chilli powder
½tsp ground turmeric
1tsp ground coriander
3 curry leaves
2tsp finely chopped fresh coriander leaves

- *Serves 4* 
- *175cals/735kjs per serving*

1 Mix and wash both dals in several changes of water, until it runs clear. Leave them to soak in fresh water for 30 minutes.

2 10 minutes before the pulses are ready, heat the butter in a saucepan over medium heat. When the fat is hot, add the cumin and mustard seeds. As soon as they pop and splutter, add the chopped onion, ginger, garlic and chilli, if wished. Stir-fry the mixture until the onion is golden, then add the salt, chilli powder, turmeric, ground coriander and curry leaves. Mix well and pour in just enough water to cover the dals by a depth of 4cm/1½in. Add tomatoes.

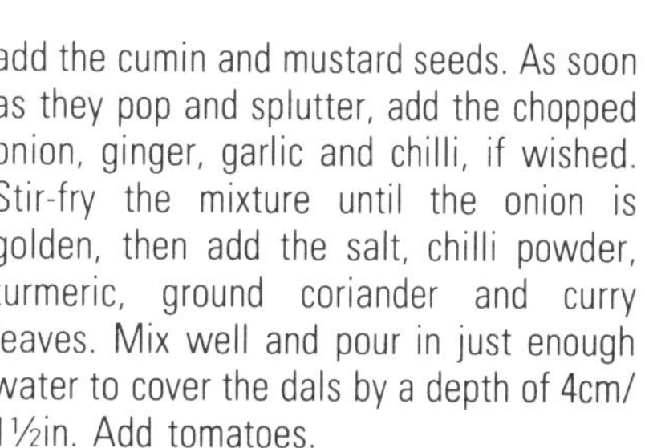

3 Cover the pan and cook over gentle heat until the dals are tender and the mixture has the consistency of thick soup. Stir to blend well. Serve hot, garnished with the coriander leaves.

Chicken curry

- *Preparation: 25 minutes*
- *Cooking: 40 minutes*

1.6kg/3½lb chicken cut into 8 pieces
2.5cm/1in piece of fresh root ginger
6 garlic cloves
4 onions
50g/2oz butter
½tsp chilli powder
4 cardamom pods, crushed
3 cloves
2.5cm/1in cinnamon stick
150ml/¼pt Greek yoghurt, whisked
2tsp white malt or wine vinegar
salt
2tsp sugar

- *Serves 4* 
- *530cals/2225kjs per serving*

1 Skin the chicken. Grate the ginger, garlic and two of the onions and reserve. Slice the remaining onions lengthways. ▶

2 Put the butter in a frying pan over medium heat and when hot, brown the sliced onions on both sides; remove half the cooked onions and reserve. Add the grated ingredients, chilli powder, cardamom, cloves, cinnamon, yoghurt, vinegar, salt, sugar and chicken pieces to the pan. Cook them until all the liquid evaporates, stirring frequently.

3 Add 100ml/3½fl oz water to the pan, bring to the boil, lower the heat and simmer until oil separates from the gravy; about 20 minutes. Sprinkle the chicken with the reserved onions and serve hot.

Cook's tips

The chicken is skinned so that spices can penetrate the meat more fully. Clarified butter would be used in Bengal because fresh butter soon goes rancid in a very hot climate.

Mango-stuffed okra

- **Preparation: 10 minutes**
- **Cooking: 30 minutes**

450g/1lb okra
2tbls mango powder (see Unusual ingredients, page 121)
1tbls ground coriander
2tsp ground cumin
½tsp chilli powder
1tsp salt
½tsp garam masala (see page 121)
3tbls oil

- **Serves 4**
- **80cals/335kjs per serving**

1 Wash the okra quickly under running water and dry them thoroughly with absorbent paper. Prolonged washing would make the cooked okra rather gluey.

2 Carefully cut off the stem end, then slit each one along its length on one side, but do not cut completely through.

3 Mix together the rest of the ingredients except the oil. Carefully open each okra slit and stuff some of the spice mixture inside. Reserve the rest.

4 Heat the oil in a shallow frying pan over medium heat. Carefully add the okra, turn the heat to medium-low and cook for 7-8 minutes.

5 Sprinkle any remaining spice mixture on top, cover, turn the heat to low and cook for about 10 minutes, stirring once or twice to prevent the okra from sticking. Serve hot.

Mango stuffed okra

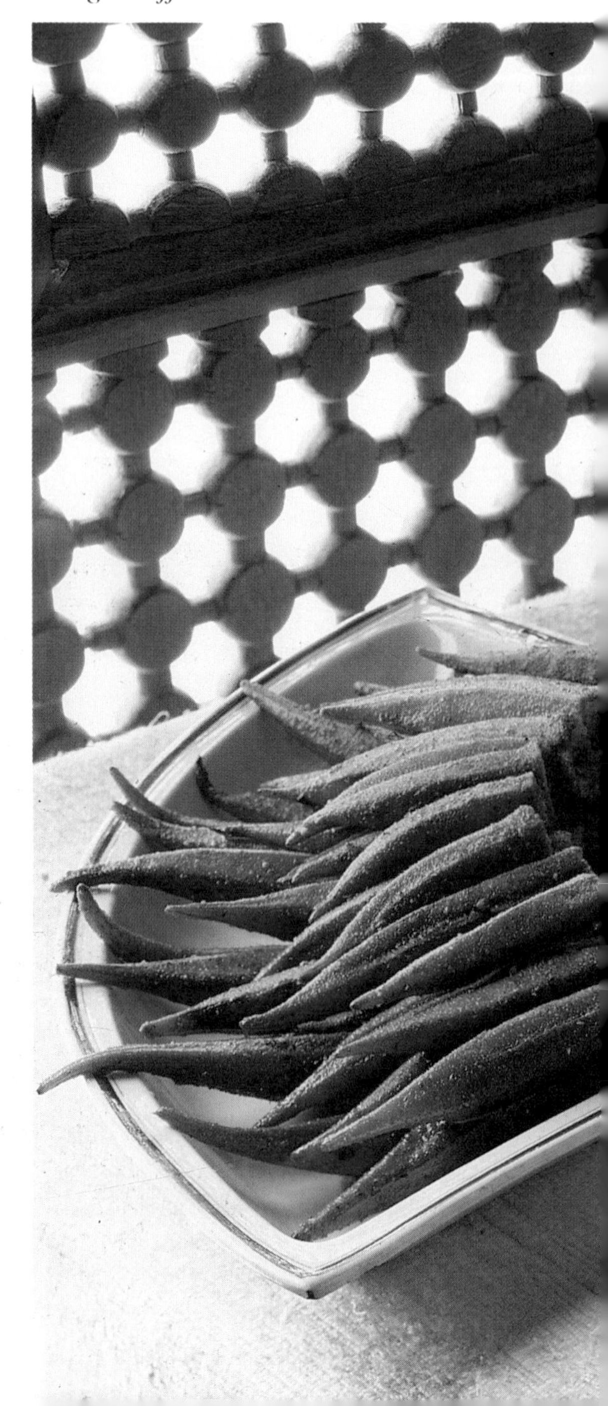

Pineapple chutney

- **Preparation: 10 minutes, plus cooking**
- **Cooking: 10 minutes**

750g/1¾lb canned pineapple chunks or rings in syrup cut into pieces
1tsp vegetable oil
¼tsp black mustard seeds
1tbls raisins
75g/3oz sugar
1tsp salt
1tsp cornflour

- **Serves 8** 
- **90cals/380kjs per serving**

1 Drain the syrup from the canned pineapple and reserve. Heat the oil in a small saucepan over low heat, and add the mustard seeds. When the seeds begin to pop, add the pineapple chunks or pieces, stir for 3 minutes and add the reserved pineapple syrup, raisins, sugar, salt and 50ml/2fl oz water.

2 When the liquid begins to simmer, add the cornflour mixed with 2tsp cold water. Stir for 1 minute, or until the liquid begins to thicken, then remove from heat. Pour the chutney into a serving dish, leave to cool, then refrigerate.

Spicy fried liver

- *Preparation: 10 minutes, plus 30 minutes marinating*
- *Cooking: 10 minutes*

450g/1lb lamb's liver
2 garlic cloves, well crushed
2 green chillies, finely chopped (optional)
½tsp chilli powder or to taste
¾tsp salt or to taste
5-6tbls lemon juice
3-4tbls vegetable oil
2tbls freshly chopped coriander leaves

- *Serves 4* 
- *445cals/1870kjs per serving*

UNUSUAL INGREDIENTS

Garam masala is a mild seasoning mixture of ground spices. It can be bought in supermarkets or Indian shops but is quite easy to make at home. Grind together 3tbls black peppercorns, 2tsp cumin seeds, 2tbls cloves, 15 green cardamon pods, 6 bay leaves and 1½tbls ground cinnamon. Store it, airtight, for up to 2 months.

Mango powder is made from ground, dried mango slices. It is available from Indian stores.

Moong dal is a type of split yellow lentil, available at Indian stores.

Dal is any kind of lentil, often used in soup, available at most supermarkets.

1 Cut the liver into bite-sized pieces. Put them in a bowl with the garlic, chillies, if using, chilli powder, salt and lemon juice. Leave the meat to marinate for 30 minutes.

2 Heat the oil in a frying pan over medium-high heat. Remove the meat from the marinade with a slotted spoon and add it to the pan. Stir-fry over high heat for 5 minutes or until the liver is cooked but still tender. Do not overcook.

3 Transfer the liver to a serving platter. Pour any remaining marinade over the liver and sprinkle with coriander leaves. Serve at once with the mixed salad and rice.

Lentil and vegetable dal

- *Preparation: 15 minutes*
- *Cooking: 1¼ hours*

250g/8oz continental lentils or split peas
6 cauliflower florets
3 radishes, cut in 1cm/½in slices
1 small courgette, cut in 2.5cm/1in slices
1tsp sugar
salt
1tbls butter
½tsp onion seeds
1 green chilli, cut in half
1 dried red chilli, slightly slit
2.5cm/1in fresh root ginger, peeled and chopped
boiled rice, to serve

- *Serves 4* 
- *225cals/1070kjs per serving*

1 Wash the lentils or split peas and place them in a saucepan. Cover with 5cm/2in water, bring to the boil and simmer for 50-60 minutes, until the pulses begin to turn tender. Top up the water regularly to keep them covered.

2 Add the vegetables and cook until everything is tender, about 10 minutes. Add the sugar and salt to taste and remove the pan from the heat.

3 Heat the butter in a frying pan over medium heat, add the onion seeds, prepared chillies and chopped ginger. Stir the spices for 1 minute. Remove the chillies, if wished, and mix the spices thoroughly into the pulses. Serve hot with plain boiled rice.

Spicy fried liver

Lentil and vegetable dal (page 121)

Sweet curd pudding

- **Preparation: 15 minutes, plus 2 hours draining**
- **Cooking: 20 minutes**

1.2L/2pt milk
200ml/7fl oz natural yoghurt, whisked
125ml/4fl oz sweetened condensed milk
1-2 drops vanilla essence
4 pistachio nuts, grated
shelled pistachio nuts, to decorate (optional)

- **Serves 8**
- **150cals/630kjs per serving**

1 Pour the milk into a saucepan and bring to the boil over medium heat. Remove from the heat, stir in 150ml/¼pt of the yoghurt and 1tbls water and stir until the milk separates completely. Place the pan over medium heat, bring to boiling point, then simmer and stir the milk for 2 minutes. Remove from the heat and strain it through a clean piece of muslin. Tie the muslin together loosely and leave the milk solids hanging until all the liquid has drained away; about 2 hours. Do not squeeze.

2 Turn out the drained milk solids and knead into a smooth paste. Stir in the rest of the yoghurt, the condensed milk and vanilla essence and spread the mixture 2cm/¾in thick on a 15cm/6in pie plate. Sprinkle with the grated nuts.

3 Put the plate in the top of a steamer or in a colander over a saucepan of gently simmering water. Steam without the lid for about 10 minutes or until the pudding is just firm. Remove the plate from the heat; allow the pudding to cool, then refrigerate it until just before serving.

4 To serve, cut the pudding into diamonds or squares and decorate each piece with a pistachio nut, if you wish.

INDIAN BREADS

Poori are a wholemeal bread speciality of northern India, served either plain with vegetable stuffings or with vegetable purée kneaded into the dough. Spinach is the most commonly used vegetable. They are deep-fried.
Chapattis are large, thin, round cakes made from wholemeal flour, with no leavening and little fat. They are cooked on a griddle.
Parathas are similar, but made with white flour and a little more fat, so that they are more flaky in texture. They are usually shallow-fried, and can be served stuffed with a savoury filling.
Nan is a slightly leavened baked bread usually torn apart and eaten instead of rice.

The Land of Spices

Pakistan and western India offer a range of cooking to tantalize the tongue and make the mouth water

Pakistani kebabs on a bed of lettuce salad (page 124)

THERE IS AN intriguing mystery to the spicy fragrance of cooking from this part of the world – it arises from the subtle way in which cooks combine their spices before they add them to the food. It is sometimes said that Pakistanis prefer the spices to the actual food and in western India the preparation of their two unique spice blends, *aadoo mirch* and *dhania zeera,* is a daily enterprise.

A lot of meat and no veg!

Traditionally, Muslims eat very few vegetables and this is typical of Pakistan too. They eat large quantities of meat, especially mutton and lamb. Pork is of course forbidden, but beef and chicken are quite popular. There are three standard ways of serving meat: korma, which is a mild, less liquid curry than its Indian counterpart; kofta, based on spicy mincemeat balls; and kebabs, pieces of meat grilled on skewers.

Rice is nice

Basmati rice is particularly fragrant and is used for the famous biriani dishes (page 126) – but since wheat is the main cereal grown in Pakistan their various types of bread are specially delicious. Nan is a leavened bread cooked in a clay oven *(tandoor),* but chapatis are unleavened and are cooked on a griddle or in a heavy iron frying pan. *Zarda* is a sweet rice dish served as a dessert, flavoured with raisins, almonds and pistachios.

Fire and spice

Move into western India and you will find fiery vindaloos and spicy ways of serving vegetables. Gujerati cooking is typical of the region: this is mainly vegetarian and a wide range of vegetables including potatoes, spinach, okra and aubergines are prepared with such spices as coriander, cumin, root ginger and garlic. The local population includes many Parsees who have blended Persian cooking with that of India to produce their most famous dish, Chicken Dhan sakh, where a savoury pulse-based curry is heavily spiced and enlivened with sharp tamarind juice. This dish is now extremely popular in restaurants in the UK.

Can this be pork?

If you are in Goa, now part of western India, your spicy hot vindaloo might well be pork, for the Goans are mainly Christians, converted by the original Portuguese colonists, and they have no pork prohibition. It is their free and easy use of chillies which makes their vandaloo so fierce. No wonder that throughout this area, cool *raita* is served with hot dishes, a simple mixture of yoghurt with cucumber (page 116).

Sweets for sticky fingers

Although sweets are sometimes made at home, it is more usual to buy them in specialist sweet shops where a colourful array of various sugary temptations lies in wait. At home, *shrikhand* (Sweet saffron yoghurt, page 128) brings the meal to a close along with honey-based puddings or cardamom-scented Carrot halva (page 116).

Pakistani kebabs

Also known as *Sheek kebabs*, these skewers are threaded with spiced cubes of meat, or with spicy minced meat shaped into thin sausages.

- ***Preparation: 15 minutes, plus 5 hours marinading***
- ***Cooking: 10-15 minutes***

450g/1lb shoulder of lamb or chuck steak, cut into 4cm/1½in cubes
1 onion
2.5cm/1in piece of fresh root ginger
2 garlic cloves
½tsp turmeric
½tsp poppy seeds
½tsp cumin seeds
6 black peppercorns
seeds of 2 cardamom pods
2 cloves
¼tsp chilli powder
¼tsp cinnamon
2tbls natural yoghurt
2tsp vegetable oil
salt
2tsp clarified butter or vegetable oil for brushing

- ***Serves 4-6***
- ***440cals/1850kjs per serving***

1 Finely grate the onion, ginger and garlic into a bowl. Grind the spices using a mortar and pestle. Mix them into the onion mixture with the chilli powder, cinnamon, yoghurt, the oil and add a little salt to taste.

2 Add the meat to the bowl and turn it in the mixture. Leave to marinate for 4-5 hours in the refrigerator.

3 Heat the grill to medium. Wipe the meat dry, thread it on to 12 metal skewers so that the pieces of meat do not touch. Cook under the grill for 10 minutes or until the meat is done, turning the skewers and basting frequently with fat. Do not overcook. Serve the kebabs immediately with salad and chapatis or spiced lentils.

Variations

For minced meat Sheek kebabs, knead 450g/1lb good quality minced lamb or beef with 1tbls each of gram flour, lemon juice and chopped coriander leaves and all the ingredients given above except for the yoghurt and oil. Brush 12 metal skewers with oil and shape the meat around the skewers into thin sausages. Cover the rack with foil and turn the skewers on it, under the grill, for about 15 minutes, brushing occasionally with oil until the meat is evenly browned, crusty and cooked inside.

Indian flat bread

Chapatis

Made from wholemeal flour, chapatis contain no raising agents and so are flat and unleavened

- ***Preparation: 10 minutes, plus 30 minutes resting***
- ***Cooking: 20 minutes***

250g/9oz wholemeal flour
1tsp salt
1tbls clarified butter

- ***Makes 6 chapatis***
- ***135cals/565kjs per chapati***

1 Sift the flour and salt together in a bowl. Make a well in the centre and add 4tbls water, a little at a time, to the centre. Using your fingers, pull the flour from the side into the liquid.

2 Knead the mixture 5-10 minutes until it forms a soft, smooth and pliable dough. Cover and leave for 30-60 minutes.

3 Knead the dough 2-3 minutes and divide it into 6 equal portions. Roll the portions into balls, dust them with flour and flatten them slightly. Roll each ball out to about 20cm/6in diameter.

4 Heat an ungreased griddle or heavy iron frying pan until hot. Gently remove the chapati from the rolling board and put it on the hot griddle. Cook for a few seconds, then turn the chapati over with a spatula or fish slice. If the first chapati sticks grease the pan very lightly with clarified butter.

5 As soon as small bubbles appear on the surface of the bread turn it again. Press the edges of the chapati down on the pan with a thick dry cloth to ensure even cooking. Continue to cook the bread, turning again. When the brown spots appear on both sides, it is ready.

6 To make the chapati puff up, press the centre before removing from the griddle (or put the chapati briefly under a hot grill).

COOL AND REFRESHING

Traditionally, no alcohol is served in any part of India, but recently a few people have begun to indulge in beer or wine. Usually people drink *lassi*, a drink made from yoghurt and flavoured with cardamom and rose water, or fruit juices.

Royal korma

Shahi korma

A mild but tasty dish

- ***Preparation: 30 minutes***
- ***Cooking: 2¼ hours***

450g/1lb very lean lamb, mutton or veal, cut into 4cm/1½in cubes
2 large onions
6 garlic cloves
2.5cm/1in piece of fresh root ginger
2tbls clarified butter
½tsp chilli powder
1tsp ground coriander
½tsp ground cumin
6 cardamom pods, crushed
400ml/14fl oz milk
1tsp sugar
1tsp poppy seeds, ground in a mortar and pestle
1tbls ground almonds
pinch of saffron
spiced rice or Chapatis (see right), to serve

- ***Serves 6***
- ***360cals/1510kjs per serving***

1 Finely grate 1 onion, the garlic and ginger onto a plate. Crush them together with the flat side of a knife to a smooth paste.

2 Slice the other onion thinly. Heat the frying pan over medium heat, fry the slices until golden brown and remove them with a slotted spoon, drain and reserve.

3 Add the onion paste to the pan with the chilli powder, coriander and cumin. Fry for 3-4 minutes, sprinkling a little water into the pan from time to time when the spices begin to stick.

4 Toss the meat and cardamom pods into the pan and season with salt. Stir the mixture for 5-10 minutes turning the meat until it is golden brown on all sides and all the liquid has evaporated.

5 Pour in the milk and sugar; simmer covered, over a very low heat until the meat is half-tender, 50-60 minutes. Alternatively to avoid sticking cook in the oven at 180C/350F/gas 4. Add the ground poppy seeds and the fried onion slices and continue cooking for about 1 hour or until the meat is tender and the liquid has thickened sufficiently.

6 Sprinkle the ground almonds and saffron, cook for 5 minutes and serve hot with Pilau rice or Chapatis. A selection of chutneys is usually provided.

Biriani

• Preparation: 35 minutes

• Cooking: 2 hours

450g/1lb leg of lamb or rump of beef, cut into 4cm/1½in cubes
5 cardamom pods
2 large black cardamom pods
5 cloves
10 black peppercorns
4 onions
5cm/2in piece of fresh root ginger
4tbls clarified butter
2 bay leaves
salt
1tsp cinnamon
450g/1lb basmati rice
2-3tsp sugar
½tsp cumin seeds
pinch of saffron
2tbls boiling water
1-2tsp red vegetable food colouring

• Serves 6-8 

• 540cals/2270kjs per serving

SPICES RARE AND WONDERFUL

European cooking was transformed when cumin, turmeric, cinnamon and ginger were brought from the East. Unusual spices can be bought and for many of them there is no substitute. Copra, dried coconut flesh, adds flavour to *zarda* (Sweet pilau, page 128); ordinary coconut is not suitable.

1 Crush together 2 of the cardamom pods, the black cardamom pods, 2 of the cloves and 5 of the peppercorns using a mortar and pestle. Finely grate 1 onion and half the ginger.

2 Heat 1tbls of the fat in a large pan over low heat and stir in the crushed spices, 1 of the bay leaves and the grated onions and ginger; continue stirring briskly for 3 minutes.

3 Add the meat to the pan, season with salt and stir continuously until the meat juices have evaporated. Add 600ml/1pt water to the pan, raise the heat and simmer until the meat is cooked, about 1 hour. Strain the stock from the meat and reserve both.

4 Crush remaining cardamom pods and cloves with the cinnamon using a mortar and pestle. Halve 1 onion lengthways and slice the halves lengthways. Chop the rest of the onions and the ginger. Wash the rice and soak it in cold water for 10 minutes.

5 Meanwhile brown the onion slices in 2tbls of the fat in a large saucepan over medium heat for 10 minutes or until crisp. Remove from the pan with a slotted spoon and drain on absorbent paper.

6 In the same fat, stir in the chopped onion, ginger, ground spices and remaining bay leaf. Continue cooking over medium heat until the onion is slightly golden, then add the strained rice. Stir for 3 minutes, then add the reserved stock, the sugar and salt to taste. If necessary, add water to make the liquid cover the rice by at least 4cm/1½in.

7 Bring the liquid to the boil, cover the pan tightly, reduce the heat and simmer the liquid. When the rice is half-cooked, about 10 minutes, add the reserved meat. Continue cooking, covered, on very low heat until all the liquid has been absorbed, about 20 minutes.

8 Meanwhile heat the rest of the fat in a small shallow pan over medium heat until it sizzles. Add the cumin seeds and the rest of the peppercorns, stir and pour the fat and spices carefully into the rice with a fork.

9 Soak the saffron strands in 2tbls boiling water for 1minute and dilute the cochineal or food colouring with 1tbls water. Stir to obtain an even colour.

10 Pour the saffron with its water and the red coloured water over the rice in 2 separate areas of the pan with white rice in between. Carefully mix the coloured areas together with two forks for a mixture of pink, yellow and white. Scatter the browned onion slices on top, cover and keep hot until serving.

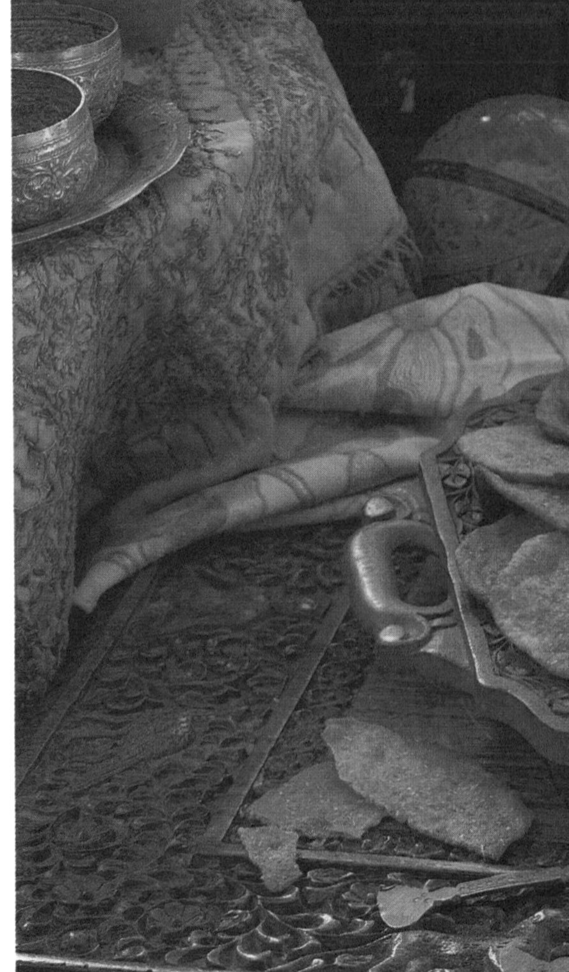

Crisp Indian bread

Kadak puri

- **Preparation: 15 minutes**
- **Cooking: 15 minutes**

400g/14oz wholemeal flour
50g/2oz gram flour
½tsp turmeric
1tsp chilli powder
pinch of salt
200ml/7fl oz milk
vegetable oil for frying

- **Makes 18-20 puris** 
- **125cals/525kjs each**

1 Place the flour, turmeric, chilli powder and salt in a bowl. Add the milk (or water) a little at a time to bind the flour into a very stiff dough: you may not need to use all the liquid.

2 Break off small pieces of dough, about the size of a walnut, and roll them out into thin rounds, no bigger than 5cm/2in in diameter and 2mm/⅛in thick. Prick the surface lightly with a fork -- this will prevent Indian breads from puffing up during frying.

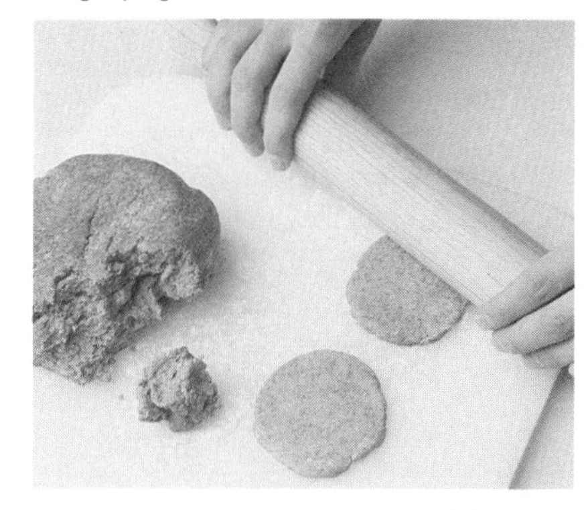

3 Place oil to a depth of 5cm/2in in a deep fat frier. Heat till nearly smoking and then carefully put a few of the rounds into the pan and fry until they are a rich golden colour.

4 Remove with a slotted spoon and drain on absorbent paper, allow to cool. Store the breads in an airtight container. They will keep crisp for up to a week. Serve as a snack or side dish.

Cauliflower and peas in chilli sauce

- **Preparation: 10 minutes**
- **Cooking: 30 minutes**

2tbls vegetable oil
1tsp mustard seeds
½tsp chilli powder
1 medium-sized cauliflower, in florets
1 medium-sized potato, cut into 10mm/½in cubes
2 medium-sized ripe tomatoes, blanched, skinned and finely chopped
½tsp turmeric
½tsp Aadoo mirch blend (see Cook's tips)
pinch of salt
1tbls chopped fresh coriander leaves
1tsp molasses
100g/4oz fresh or frozen peas
Crisp Indian bread or boiled rice to serve

- **Serves 4** 
- **370cals/1555kjs per serving**

1 Heat the oil in a heavy-bottomed saucepan over medium-high heat, add the mustard seeds and as soon as they pop add the chilli powder. Shake the pan for a second, then add the cauliflower florets and toss round in the oil.

2 Stir-fry for a few seconds, then add the potatoes, tomatoes, turmeric, Aadoo mirch, salt, coriander leaves and the molasses stirring birskly.

3 Stir well, cover and cook for 3-4 minutes, then add 425ml/15fl oz water, mixing thoroughly. Reduce the heat, cover and cook until the vegetables are tender, and the sauce has thickened slightly, about 20 minutes. (The potatoes will help to thicken the sauce.) Add the peas and cook for 10 minutes. Serve hot with chapatis or boiled rice.

Cook's tips

To make 75g/3oz of the Aadoo mirch spice blend: peel and chop 50g/2oz fresh root ginger and 1 clove of garlic. Remove stalks from the chillies. Place them all in a blender with ½tsp salt, and purée until smooth. Store in the refrigerator for up to 1 week.

Sweet saffron yoghurt

Shrikhand

- ***Preparation: 30 minutes, plus 2 hours chilling***
- ***Cooking: 10 minutes***

600ml/1pt yoghurt
350g/12oz caster sugar
a few strands of saffron
1tbls milk
pinch freshly grated nutmeg
fresh fruit to serve (optional)

- ***Serves 6*** 
- ***275cals/1155kjs per serving***

1 Tie up the yoghurt in a cheesecloth and place over a bowl for 5 minutes to allow the liquid to drain.

2 Place the drained yoghurt in a bowl, add the sugar and mix thoroughly until the sugar has completely dissolved.

3 Crush the strands of saffron using a mortar and pestle. Put the milk in a small saucepan and place over a low heat. When the milk is warm but not hot add the saffron powder and stand for a few minutes to infuse.

4 Add the sweetened yoghurt together with the nutmeg. Stir thoroughly and then chill for 2 hours. Traditionally this Maharashtran dish is served with puris but it is also very good with fresh fruit.

Sweet pilau

Zarda

This colourful, sweet rice dish is always served as a dessert. The pistachio nuts and almonds add a delightful contrast in texture to the soft rice

- ***Preparation: 30 minutes, plus 6 hours cooling***
- ***Cooking: 1 hour***

350g/12oz basmati rice
small piece of unpeeled copra, cut into thin, 5cm/2in slices

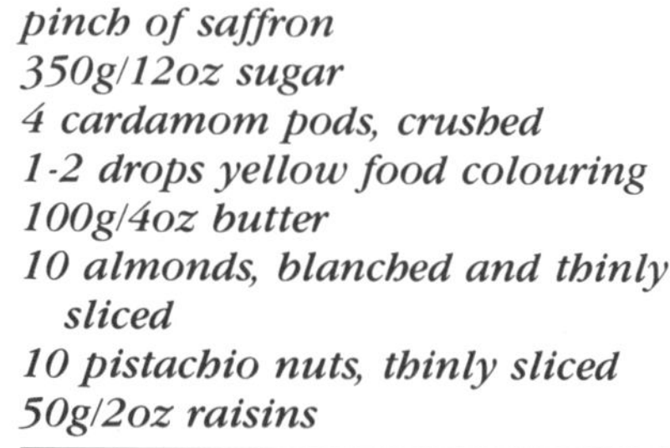

pinch of saffron
350g/12oz sugar
4 cardamom pods, crushed
1-2 drops yellow food colouring
100g/4oz butter
10 almonds, blanched and thinly sliced
10 pistachio nuts, thinly sliced
50g/2oz raisins

- ***Serves 6***  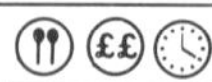
- ***610cals/2500kjs per serving***

1 Wash the rice thoroughly, drain and place in a large saucepan with 700ml/1¼pt water. Cook the rice over medium heat for 8-10 minutes, until it is slightly under-cooked. Drain the rice well and spread it over a large platter or baking tray to dry while preparing the recipe.

2 Wash the copra slices and soak them in cold water for 20 minutes. Soak the saffron in 1tbls hot water for about 10 minutes, drain and reserve the liquid.

3 Meanwhile mix the sugar with 600ml/1pt water in a saucepan, add the crushed cardamoms and 1-2 drops of yellow food colouring. Boil the mixture over medium-heat until a drop of the syrup feels sticky on a plate, this will take about 15-20 minutes.

4 Stir the rice into the syrup, add the butter and continue cooking over low heat. Drain the copra slices and stir them into the syrup, together with the almonds, pistachios and raisins. Continue cooking until the syrup is completely absorbed by the rice. Add the saffron liquid and stir in carefully with a fork until the rice is uniformly coloured. Leave the Sweet pilau until cold before serving.

Dining Down Under

All too often associated with barbecues, beer and billy tea, Australia and New Zealand have gently evolved their own cuisine, full of ethnic sparkle

WHEN THE FIRST white settlers came to the south-eastern coast of Australia in 1788, they found a land warmer than their native England and only strange animals, and even stranger plants. Not unreasonably, they kept to familiar food such as meat from their imported livestock and grains and fruit raised from seeds sent from England. Even today, although there are some unique dishes, much Australian food is English in style, including plum pudding served for midsummer Christmas.

The first settlers in Australia called their colony New South Wales. Bit by bit, other parts of the coast were settled and Queensland in the warmer north-east added delicious tropical fruit and vegetables, such as passion fruit, avocados, pineapples, dates, figs, bananas and chokos (small squashes, also known as chayotes) to the Australian diet.

A melting-pot of people

With the discovery of gold in the outback in 1851, suddenly thousands of English and Scottish immigrants, followed by Americans, Germans, Chinese and others, came to Australia and the population tripled within a few years. Australians began to enjoy a much wider range of cooking styles as the immigrants' cuisines became known and popular. Today Australia's major cities boast excellent restaurants serving foods of all nations (Chinese, Italian, Indian, Japanese and Thai, to name just a few) as well as fine Australian wines.

Pavlova (page 132)

Sheep grazing in a meadow in Australia

Most of the Australian population lives around the coast and seafood features in many of the national dishes such as baked barramundi and carpetbag steak (beef rump stuffed with oysters). Australian murray cod is highly esteemed for its delicate flavour, and Sydney rock oysters are among the world's best.

Having bred exceptionally high-quality livestock, especially sheep and cattle, both Australians and New Zealanders consume large quantities of meat – the national institution, the barbecue, features grilled beef or mutton without fail. But rabbit and chicken are just as popular, often cooked with exotic fruit.

Land of lamb

New Zealand, being further south, is cooler and – unlike parts of Australia – it has reliable rainfall throughout the country. Two-thirds of the area is grassland and clover, which feeds enormous flocks of sheep and has turned New Zealand into the world's largest exporter of lamb, mutton and dairy produce. Known for their healthy appetite, New Zealanders eat 110kg/243lb meat each a year, and most of it is lamb.

The most typical national dish, called 'colonial goose', is a roast shoulder of lamb with breadcrumb stuffing. They also claim Pavlova (page 132), and the Chinese gooseberry, or kiwi fruit, which is now exported all over the world.

New Zealand's other fruit crops are marvellously varied: peaches, nectarines, plums and cherries, along with subtropical fruits from the northernmost of the islands.

(AB)ORIGINAL DELIGHTS

The way of life of the Aborigines, Australia's original inhabitants, was closely related to the natural environment upon which they depended for food. Their culinary traditions include kangaroo meat, cockchafer grubs, bats and lizards. Kangaroo-tail soup is considered a special delicacy, but kangaroos are now a protected species.

'CHINESE' GOOSEBERRY

The kiwi fruit is about the size of a large egg, with a greenish-brown hairy skin and pale green, highly perfumed and juicy flesh. It is grown largely in New Zealand and can be used in various ways: served fresh and halved as a dessert, peeled and cut into cubes or slices for fruit salads, or as a colourful garnish for savoury dishes such as baked mackerel or fried pork chops. It is rich in vitamin C.

Australian barbecued beef

- ***Preparation: 15 minutes, plus overnight marinating***
- ***Cooking: 10 minutes***

1 onion, chopped
1tbls finely chopped fresh root ginger
3 garlic cloves, roughly chopped
1tbls lemon juice
1tsp ground cumin
1tsp salt
½tsp pepper
1kg/2¼lb rump steak, 2cm/¾in thick, cut into 6-8 steaks
flat-leaved parsley sprigs and tomato wedges, to garnish

- ***Serves 6-8***
- ***285cals/1195kjs per serving***

1 Place the chopped onion, ginger, garlic, lemon juice, ground cumin, salt and pepper in a blender and blend to a paste.

2 Rub the marinade onto the steaks, cover with stretch wrap and chill overnight.

3 Remove the steaks from the fridge to allow them to come to room temperature while you prepare the barbecue or heat the grill to high.

4 Grill the steaks for 3-4 minutes on each side, or until done to your liking. Serve at once, garnished with flat-leaved parsley and tomato wedges.

Australian garlic prawns

- ***Preparation: 20 minutes, plus overnight marinating***
- ***Cooking: 15 minutes***

10 garlic cloves, crushed
250ml/9fl oz corn oil
250ml/9fl oz dry white wine
2tbls lemon juice
salt and pepper
4tbls chopped parsley
1kg/2¼lb large raw prawns, heads removed but not shelled
crusty bread, to serve

- ***Serves 6-8*** 
- ***560cals/2350kjs per serving***

1 Crush the garlic and mix with the oil, wine, lemon juice, 1tsp salt, pepper to taste and the parsley. Leave overnight for the flavours to blend.

2 One hour before cooking add the prawns, in their shells, to the marinade and coat them well.

3 Prepare the barbecue or heat a flameproof casserole over moderate heat. Add the prawns with their marinade if cooking indoors, cover and cook for 15 minutes. If cooking on the barbecue, baste with the marinade while cooking. Serve the prawns with bread to soak up the sauce.

SWEET POTATOES

The orange-glazed chops can be served with sweet potatoes. Boil two sweet potatoes in their skins for 1 hour or until tender. Peel and slice thickly. Brush with melted butter and grill under a medium grill until lightly browned on both sides, brushing occasionally with melted butter.

IN THE BASKET

For a spectacular dessert after the barbecue, why not serve a deliciously refreshing Australian melon basket? It should contain as many fruits as possible from the Queensland 'fruit paradise' – choose from melons, passion fruit, kiwi fruit, strawberries, bananas and pineapple. Finish off by sprinkling passion fruit seeds on top.

Orange-glazed lamb chops

- ***Preparation: 20 minutes***
- ***Cooking: 20 minutes***

1 large orange
2 kiwi fruit
4 large loin lamb chops, about 2cm/¾in thick
2tsp oil
1 onion
1tbls clear honey
strips of orange zest, to garnish

- ***Serves 4*** 
- ***270cals/1135kjs per serving***

1 Grate the zest from half the orange, then cut the orange in two along the grated half. Squeeze the juice from the grated half and reserve the juice and zest separately. Cut the second half into four crescents and reserve. Peel the kiwi fruit and cut them into four slices each.

2 Snip around the outer edges of the chops in several places to prevent them curling. Brush the bottom of a frying pan with the oil. Put in the chops, then fry them over medium-high heat for 6-8 minutes until they are browned on both sides and cooked to your liking.

3 Remove the frying pan from the heat, put the chops on a warm plate and keep warm. Grate the onion into the frying pan and brown lightly. Add the honey to the pan and stir until it begins to darken slightly. Add the reserved orange zest and juice and stir until syrupy.

4 Add the chops, then the orange and kiwi slices, to the pan and turn them over in the glaze until they are coated. Arrange them on a warmed serving dish.

5 Pour the remaining glaze over the chops, garnish and serve immediately.

Anzac biscuits

- ***Preparation: 20 minutes***
- ***Cooking: 25 minutes***

100g/4oz flour
200g/7oz Demerara sugar
75g/3oz coarse oatmeal
50g/2oz desiccated coconut
100g/4oz butter, plus extra for greasing
1tbls golden syrup
½tbls bicarbonate of soda
2tbls boiling water

- ***Makes 25***
- ***85cals/355kjs per serving***

1 Sift the flour into a large bowl. Add the sugar, oatmeal and desiccated coconut and mix well.

2 Melt the butter with the syrup in a small saucepan over low heat.

3 Stir the bicarbonate of soda into the boiling water until dissolved, then add to the butter and syrup. Add the liquid mixture to the dry ingredients and blend together thoroughly. Chill for 10 minutes. Heat the oven to 170C/325F/gas 3 and grease two large baking sheets.

4 Put 2tsp of the mixture at a time on the baking sheets, spacing well apart. Flatten into small rounds and bake for 16-20 minutes or until golden brown.

5 Cool the biscuits on the sheets for 10-15 minutes, then transfer them to a wire rack until cold.

Pavlova

- ***Preparation: 1¼ hours***
- ***Cooking: 1 hour, plus cooling***

4 large egg whites
pinch of salt
225g/8oz caster sugar
1tsp white wine vinegar
½tsp vanilla essence
1½tsp cornflour
2 nectarines or peaches
100g/4oz small strawberries, halved
1 kiwi fruit, peeled and thinly sliced
600ml/1pt double cream
3tbls icing sugar
fresh mint leaves, to decorate

- ***Serves 8*** 
- ***440cals/1850kjs per serving***

1 Heat the oven to 130C/250F/gas ½. Line two baking sheets with non-stick baking paper and mark two circles, one about 15cm/6in diameter, the other 10cm/4in diameter, on the paper.

2 In a large, clean, dry bowl, whisk the egg whites with the salt until stiff. Add the caster sugar, 1tbls at a time and whisking between each addition, until the meringue is stiff and glossy. Whisk in the vinegar, vanilla essence and cornflour.

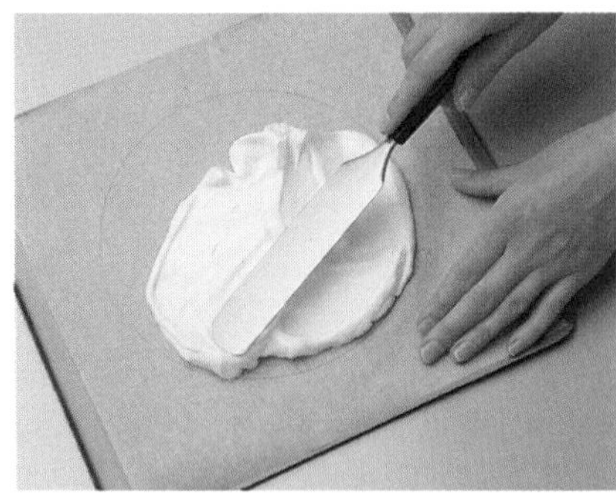

3 Spread each circle with meringue, 1cm/½in thick, working outwards to the line using a palette knife. Reserve one quarter of the meringue.

4 Fit a piping bag with a large star nozzle and spoon in the remaining meringue. Pipe a ring of meringue around the outside of each circle on the paper. Pipe a second ring outside the first.

5 Bake the meringues for 1 hour until very lightly coloured. Loosen the meringues carefully from the paper and leave to cool completely in the turned-off oven with the door ajar for 1 hour.

6 Pour boiling water over the peaches, if using, and wait for 10 seconds, then remove and peel. Halve the nectarines or peaches, remove the stones; slice thinly.

7 Carefully transfer the larger meringue circle to a serving plate. Whip the cream with the icing sugar until stiff, then spread half in the centre of the larger meringue. Arrange the nectarine or peach slices and strawberry halves on the cream, reserving 8 halves for decoration.

8 Put on the second meringue layer, then carefully spread most of the remaining cream in the centre. Pipe the rest of the cream into six rosettes around the pavlova. Put two strawberry halves in the centre, surround with slices of kiwi fruit and decorate with mint leaves, then put strawberry halves and kiwi fruit slices around the cream rosettes. Serve as soon as possible.

DANCER'S TUTU

Pavlova, the classic New Zealand dessert, was created for the Russian ballerina Anna Pavlova to honour her most famous role as the dying swan. The meringue, which spreads slightly during cooking, is supposed to resemble a dancer's tutu.

Islands in the Sun

Based on root vegetables, rice, beans and fruit, West Indian cooking is a melting pot of traditions and food from every corner of the globe

Piña colada (with stirrers), Planter's punch and Banana chips (all page 136)

THE ISLANDS OF the Caribbean stretch in a sparkling chain from Florida to Venezuela, with a cuisine as varied as their geographical span. Many countries have played a part in the development of the islands' cooking. Columbus and other Spanish explorers introduced the food of Europe, Asia and Africa, bringing in cattle, sheep, goats, pigs, chickens, rice, cooking oil and sugar cane. After the Spanish conquest of Mexico, tomatoes, avocados, chocolate and turkeys reached the islands from the American mainland. Africans brought foods such as okra and black-eyed peas, while labourers from India brought curry spices and mangoes to the islands.

Taking roots

Nowhere are root vegetables more highly esteemed. Yams are edible tubers like potatoes, but they come in many sizes and shapes. They have a slightly nut-like flavour and a texture similar to potatoes. There are also sweet potatoes with their mealy, dry texture that welcomes butter, gravy or sauces.

Cassava is the white, starchy root from which tapioca is made. It is boiled and eaten as a vegetable, or ground to make cassava meal for bread and cakes.

Seafood

There were few animals when Columbus arrived, but there were, and still are, abundant fish and shellfish. Even now island cooking is not

1 Remove the seeds from the pumpkin and cut into slices. Peel the skin off each slice and cut the flesh into chunks.

2 Heat the butter in a saucepan and fry the onion until soft but not browned, stirring occasionally. Add the garlic, pumpkin, tomatoes and stock and simmer, covered, for about 30 minutes or until the pumpkin is tender.

3 Cool the mixture slightly and push it through a sieve or reduce to a coarse purée in a blender or food processor.

4 Return the mixture to the pan, season with salt and pepper to taste, add the cream and hot pepper sauce and reheat gently.

Cook's tips

Pumpkin is usually only available in the autumn, although you might find it in ethnic shops at other times of the year.

Freezer

Pumpkin will keep for 1 year in the freezer: prepare, then steam or boil until tender. Mash, cool, then freeze in rigid containers.

Carrying a heavy load on your head is made to look easy by these Jamaican women selling bananas

based on meat and poultry. The emphasis is on vegetables, rice, beans and fresh fruit such as pineapples, bananas, plantains, guavas, papayas, mangoes and watermelons, which are eaten at breakfast, lunch and often inbetween times.

Pumpkin soup

- ***Preparation: 30 minutes***
- ***Cooking: 40 minutes***

1kg/2¼lb pumpkin
2tbls butter
450g/1lb onions, finely chopped
1 garlic clove, chopped
2 tomatoes, skinned and chopped
1L/1¾pt chicken stock
salt and pepper
250ml/9fl oz single cream
few drops of hot pepper sauce

- ***Serves 6-8***
- ***145cals/610kjs per serving***

CREAMED COCONUT

Coconut gives many West Indian dishes their characteristic richness. The easiest way is to use a block of creamed, unsweetened coconut which you'll find in supermarkets or ethnic grocers. It will keep almost indefinitely in the refrigerator if well wrapped.

Rice and peas

- ***Preparation: 15 minutes, plus overnight soaking***
- ***Cooking: 1 hour 5 minutes***

200g/7oz red kidney beans, soaked overnight
2tbls oil
1 onion, finely chopped
100g/4oz creamed coconut
1 fresh chilli, seeded and chopped
1 dried red chilli, seeded and chopped
½tsp dried thyme
salt and pepper
450g/1lb long-grain rice

- ***Serves 6-8*** 
- ***470cals/1975kjs per serving***

1 Drain the beans and rinse well. Place in a large saucepan or casserole with enough water to cover by 5cm/2in and bring to the boil. Boil rapidly for 10 minutes, then simmer for 30 minutes or until the beans are tender, adding more hot water as necessary. Drain the beans, reserving the liquid, and return them to the saucepan.

2 Heat the oil in a frying pan and fry the onion until golden, stirring occasionally. Dissolve the creamed coconut in enough water to make 450ml/16fl oz. Add this to the beans in the pan with the fried onion, chillies, thyme, salt and pepper to taste and the rice.

3 Make up the quantity of bean liquid with cold water to 900ml/1pt 12fl oz. Add this to the saucepan and bring to the boil. Cover and simmer over very low heat for 20 minutes or until the rice is tender and all the liquid absorbed. Serve.

Cook's tips

This is traditionally made with dried pigeon peas (hence its name), but other dried beans such as haricot or black beans can also be used.

Caribbean chicken casserole

This delicious dish is typically hybrid, with strong Spanish and Mexican influences

- ***Preparation: 40 minutes***
- ***Cooking: 1¼ hours***

1.5kg/3¼lb chicken, cut into serving pieces
salt and pepper
4tbls oil
1 onion, finely chopped
1 red pepper, seeded and chopped
1 garlic clove, chopped
50g/2oz chopped ham
225g/8oz tomatoes, peeled and chopped
10 pimiento-stuffed green olives, halved
2tbls raisins
1tbls capers
1 bay leaf
2tbls tomato purée
1tbls red wine vinegar
250ml/9fl oz chicken stock or water
450g/1lb potatoes, peeled and quartered
450g/1lb frozen peas, defrosted

- ***Serves 6*** 
- ***445cals/1870kjs per serving***

1 Season the chicken pieces with salt and pepper. Heat the oil in a heavy casserole and fry the chicken pieces, turning them often, until they are golden on all sides. Lift out the chicken and reserve.

2 In the oil remaining in the casserole, fry the onion and pepper until the onion is soft but not browned, stirring often. Add the garlic, ham, tomatoes, olives, raisins, capers, bay leaf, tomato purée, vinegar and stock or water. Stir to mix, bring to the boil, then simmer for 2-3 minutes.

3 Return the chicken pieces to the casserole, cover and simmer over low heat for 35 minutes. Add the potatoes and simmer for 20 minutes more or until the chicken and potatoes are tender, adding the peas five minutes before serving. Serve hot.

Variations

If you like, you can replace 100ml/3½fl oz of the stock or water with dry sherry to make a richer gravy.

GOING BANANAS

Fried Banana chips make a delicately flavoured snack to serve with drinks. To prepare them, peel six unripe bananas and slice them across as thinly as possible. Place in a bowl of salted iced water for 30 minutes. Heat oil in a deep-fat fryer to 190C/375F or until a cube of bread browns in 50 seconds. Drain and dry the slices on absorbent paper and fry in batches until lightly browned. Drain on absorbent paper, sprinkle with salt and serve.

Mango fool

- **Preparation: 15 minutes, plus chilling**
- **Cooking: 20 minutes**

5 medium-sized ripe mangoes or 3 large ones
2-3tbls lemon juice
175g/6oz sugar
2 large eggs, lightly beaten
large pinch of salt
425ml/3/4pt single cream

- **Serves 4-6** 
- **555cals/2330kjs per serving**

1 Peel the mangoes, remove the stone and chop the flesh coarsely. Purée it in a blender or food processor with the lemon juice and half the sugar. Push the purée through a nylon sieve to remove any strings and chill thoroughly.

2 Combine the eggs, remaining sugar and the salt in the top pan of a double boiler. In a small saucepan, heat the cream until bubbles form around the edge. Slowly pour the cream into the egg mixture, stirring constantly.

3 Place the mixture over hot but not boiling water and cook, stirring constantly, for 10-15 minutes or until the mixture coats the back of a wooden spoon. Do not overcook or allow to boil. Pour the custard into a bowl and chill.

4 Very lightly fold the chilled mango purée into the custard; the mixture should have streaks of custard and mango. Serve cold, in sundae glasses.

Variations

If fresh mangoes are hard to find, use canned ones instead, and add sugar to taste. You could also try substituting other fruit such as papaya or pineapple.

Piña colada

- **Preparation: 10 minutes**

225ml/8fl oz pineapple juice
100g/4oz creamed coconut
100ml/3½fl oz golden rum
225ml/8fl oz crushed ice
For the garnish:
2 pieces of pineapple
2 maraschino cherries

- **Serves 2** 
- **315cals/1325kjs per serving**

1 Put the pineapple juice, creamed coconut, rum and crushed ice in a blender and blend for a few seconds. Pour unstrained into two tumblers.

2 Decorate each tumbler with a pineapple piece and a cherry on a cocktail stick.

Planter's punch

- **Preparation: 15 minutes**

3tbls fresh lime juice
5tsp caster sugar
100ml/3½fl oz dark rum
1tsp angostura bitters
225ml/8fl oz crushed ice
ice cubes
carbonated water
For the garnish:
2 pieces of pineapple
2 orange slices
2 maraschino cherries

- **Serves 2** 
- **245cals/1030kjs per serving**

1 Combine the lime juice, sugar, rum, bitters and crushed ice in a cocktail shaker and shake vigorously.

2 Pour the mixture, unstrained, into two tumblers and add ice cubes and carbonated water to taste. Garnish each drink with a pineapple piece, an orange slice and a cherry.

Cook's tips

To make crushed ice, place ice cubes in a strong plastic bag and beat with a rolling pin or hammer.

Pacific Paradise

Though separated by thousands of miles, the Philippines and the Hawaiian islands share a taste for unusual fruits and flavours.

Philippine prawn and pork pancake (page 138)

ISLANDS ALL OVER the world show the influence of visitors from other lands, traders, sailors and settlers, all anxious to maintain their own cookery customs whilst making the most of the local produce and cuisine.

Spanish galleons

The Malays were the first settlers in the island group we now call the Philippines, bringing with them a heavy reliance on coconut milk, but it was the first Spaniards who brought the chillis and garlic which give Filipino cooking its rich and spicy flavours. The Spanish also imported pigs, as well as a favourite method of cooking called *adobo,* a long slow simmer in vinegar and spices which both tenderised and preserved the meat. Their ships were soon loaded with produce from the New World – tomatoes, maize, avocados and coffee – which were cultivated on the islands and absorbed into native Filipino cooking.

Salted and sour

Filipinos are very fond of vinegar-based sauces – hot chilli vinegar with crushed garlic is spooned over roasts and grills. The sour sharpness of tart fruits such as tamarind and the green mango or calamansi are used in a cooking method called *sinigang* – meat and vegetables are boiled in a light sour broth. The combined flavours of salt and fish produce two very popular Filipino condiments: *bagoong,* a fish paste, and *patis,* a fishy liquid made from dried salty shrimps or anchovies. They enliven the blandness of the boiled rice which forms the basis of all local cooking.

Leis, luaus and lomi-lomi

The Hawaiian islands appeared in the Pacific after a series of volcanic eruptions; the first settlers sailed in from Polynesia loaded with products which were to become the basics of the Hawaiian diet: taro, pigs and coconuts. They were followed by Chinese and Japanese traders and the islands soon had a cooking style which was a wonderful, tasty amalgam of Oriental styles and flavours. But in spite of soy sauce, seaweed and stir-fry, the coconut remains supreme: chicken is simmered in coconut milk, grated coconut is sprinkled over casseroles and pies and the hollowed-out nut is used as a serving dish. And as for the pineapples . . . juicy and fragrant, they appear lusciously in fruit cups and salads.

Farmer taking produce to the feast of San Isidro in the Philippines

Philippine prawn and pork pancake

- **Preparation: 30 minutes**
- **Cooking: 1 hour**

2tbls vegetable oil
2 large garlic cloves, finely chopped
100g/4oz onion, finely chopped
100g/4oz minced pork
100g/4oz turnip, peeled and grated
200g/7oz palm hearts (400g/14oz can, drained), sliced
175g/6oz raw prawns, peeled and halved if large
dash of soy sauce
1 large lettuce
green stalks of 2-3 spring onions, cut in half lengthways
3 garlic cloves, finely chopped, to serve (optional)

For the pancakes:
3 large eggs, separated
75g/3oz cornflour
1/4tsp salt
vegetable oil for frying

For the sauce
3tbls sugar
1/4tsp salt
2tbls thin soy sauce
1tbls cornflour

To garnish:
Spring onion flowers (see Cook's tips)
cooked, unpeeled prawns

- **Makes about 14 stuffed pancakes** 
- **125cals/525kjs per serving**

1 To prepare the filling, heat the oil in a large saucepan over medium-low heat. When hot, brown the garlic, then add the onions. Sauté until soft, about 3 minutes. Add the pork and stir for a further 3 minutes.

2 Add the pork, turnip, palm hearts and prawns. Sauté until the vegetables and prawns are just cooked, about 5 minutes. Add soy sauce to taste.

3 Prepare the pancakes by beating the whites until they are frothy. Blend in the yolks. Dissolve the cornflour and salt in 200ml/7fl oz water and add this to the eggs. Blend well and allow to stand for at least 10 minutes.

4 Heat an 18cm/7in frying pan over low heat. When hot, brush with oil. Spread 2tbls batter in an arc around the pan, then tilt the pan to spread the batter. Cook on one side only for 30 seconds or until it can be flicked easily.

5 Transfer the pancake to a plate and reserve. If the batter becomes too thick to spread easily, stir in a little more water. Continue, brushing the pan with oil between pancakes, until all the batter is used. Separate the cooked pancakes with sheets of greaseproof paper.

6 To prepare the sauce, combine the sugar, salt, and soy sauce with 250ml/9fl oz water and bring it to the boil.

7 Dissolve the cornflour in 2tbls water and stir this into the boiling liquid, stirring until thickened, about 3 minutes.

8 To serve, put a lettuce leaf on each pancake, then 2 heaped tablespoons of filling along one edge. Put 2-3 pieces of spring onion green on top and roll up from the filling. Reheat in a hot oven for a few minutes. Transfer to individual serving dishes or a serving platter and garnish.

9 Serve the pancakes at room temperature, either with the sauce drizzled on top or passed separately in a sauce-boat. The chopped garlic can be served in a small dish as an additional topping for real garlic lovers.

Cook's tips

To make spring onion flowers, cut a 5cm/2in piece of spring onion. Shred one end, then place in iced water to curl.

How to ... prepare a pineapple

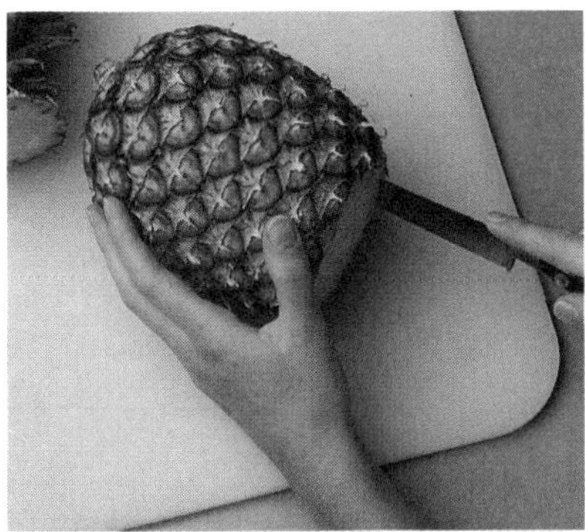

Cut the top and bottom away with a sharp knife; retain. Cut the inside to release the flesh.

With the thumbs, push the inner core out from bottom to top. Trim diagonally to remove hard skin.

Cut the flesh into rings, remove hard centre from slices. Rebuild pineapple shell to use as a container.

Pineapple coconut chicken

- ***Preparation: 30 minutes***
- ***Cooking: 25 minutes***

1.4kg/3lb chicken, boned, skinned and cut into large dice
50g/2oz butter
2 medium onions, chopped
2 medium green peppers, chopped
3tbls flour
600ml/1pt thin coconut milk (see page 140)
50g/2oz desiccated coconut
300ml/½pt chicken stock, plus extra for thinning
4 thin slices of fresh root ginger
400g/14oz fresh pineapple chunks (see above)
boiled rice, to serve (optional)

- ***Serves 4***

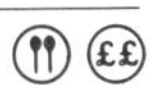

- ***600cals/2520kjs per serving***

1 Heat the butter in a large saucepan over a low heat and sauté the onions until they are translucent. Add the green pepper and cook for another 5 minutes before stirring in the flour.

2 Stir the coconut milk, coconut, chicken stock and ginger into the flour mixture. Add the chicken to the liquid and simmer, uncovered, over medium heat until the chicken is tender and the sauce has slightly thickened, about 12 minutes or until cooked. If necessary, thin with extra chicken stock.

3 Stir in the pineapple (fresh or canned), adjust the seasoning, remove the ginger slices and serve immediately accompanied by boiled rice, if wished.

FATAL ENCOUNTERS

It is curious that Ferdinand Magellan who led the first expedition to the Philippines and Captain Cook who discovered the Hawaiian Islands were both killed by native warriors before they could complete their journeys of discovery. The Spanish were quick to take over the Philippines and thanks to them most of the country is predominantly Christian; the southern islands are Muslim, though, and their cooking reflects this.

Prawns and vegetables in coconut milk

This Malayan-inspired dish is very popular in the southern Philippines. Like many of the regional specialities, this one uses coconut and chillis and is served with plain boiled rice

- ***Preparation: 20 minutes, plus making coconut milk***
- ***Cooking: 30 minutes***

2tbls oil ▶

1 large garlic clove, finely chopped
1 onion, thinly sliced
1tbls finely chopped fresh root ginger
500g/18oz aubergine, peeled and diced
250g/9oz green beans, trimmed and halved across
2tbls (or more) shrimp paste
250ml/9fl oz thick coconut milk (see below)
250g/9oz small, raw prawns, shelled
1 small chilli (optional)
plain boiled rice, to serve

• Serves 4 

• 430cals/1805kjs per serving

1 Heat the oil in a large saucepan over medium-low heat. Brown the garlic, then add the onion and cook until soft, about 3 minutes. Add the ginger, diced aubergine and green beans and cook a further 3 minutes, stirring constantly.

2 Add the shrimp paste and 150ml/5fl oz of the coconut milk. Cover and simmer until the vegetables are tender, about 15-20 minutes.

3 Add the remaining coconut milk, the prawns and chilli, if wished. Simmer until the prawns are cooked through, about 1 minute. Adjust the seasonings and serve with plain boiled rice.

Making thick and thin coconut milk

For thick coconut milk, coarsely grate fresh coconut, discarding the brown skin. Squeeze in damp muslin to remove the liquid – 1 coconut will give about 225ml/8fl oz. For thin coconut milk, add 225ml/8fl oz boiling water to the squeezed coconut, stand for 30 minutes, then squeeze in muslin as before. Repeat this process, then mix the two liquids together.

SOURCES OF SUPPLY

Many of the ingredients found in Filipino cookery can be bought in Chinese and Far Eastern specialist foodshops: shrimp paste and fish sauce (sold under the Thai name of *nam pla* in Chinese stores) are easy to find nowadays and bitter gourd, also called bitter melon, can be bought both fresh and canned.

Chicken and pork adobo

The original Spanish technique of curing a loin of pork for weeks in vinegar, olive oil and spices gave way to a much less complicated version in the Philippines, *adobo*, now the national dish

• Preparation: 25 minutes, plus 1 hour marinating

• Cooking: 1¼ hours

1kg/2¼lb chicken, cut into serving pieces
500g/18oz lean pork, cut into 2.5cm/1in cubes
100ml/3½fl oz mild wine vinegar or cider vinegar
2tbls soy sauce
3 large garlic cloves, crushed with the side of a knife
1 bay leaf, plus extra, to garnish
1tbls sugar
½tsp black peppercorns, coarsely crushed
salt
2-3tbls vegetable oil
plain boiled rice, to serve

• Serves 4 

• 660cals/2770kjs per serving

1 Put the chicken and the pork in a large stainless steel saucepan or flameproof enamelled casserole. Combine the remaining ingredients except the oil with 150ml/5fl oz water in a bowl. Pour the mixture over the meat and let it marinate for 1 hour, turning every 15 minutes.

2 Bring the liquid to the boil, cover and simmer, over low heat until the meat is tender, turning occasionally. This will take about 1 hour.

3 Remove the meat and garlic with a slotted spoon. Drain, then pat dry on absorbent paper. Strain and reserve the cooking liquid.

4 Heat the oil in a large frying pan over medium heat. Brown the garlic, then discard it. Brown the chicken and pork, a few pieces at a time and transfer them to a bed of rice on a heated serving dish. Garnish with bay leaves and keep warm.

5 Pour any remaining grease out of the frying pan and add the reserved cooking liquid. Boil for 2-3 minutes, scraping up all of the browned bits sticking to the pan. Pour the sauce over the meat or into a warmed sauce-boat and serve.

Cook's tips

For a stronger flavour marinate the meat for several hours or overnight.

Variation

This recipe can also be made with fillet of lamb in place of pork.

Vegetable, pork and shrimp paste stew

- ***Preparation: 25 minutes***
- ***Cooking: 1 hour 10 minutes***

2tbls olive oil
3 large garlic cloves, crushed
300g/11oz onion, thinly sliced
1cm/½in slice of fresh root ginger, peeled and finely chopped
500g/18oz aubergines, peeled and cut into 2.5cm/1in cubes
3 tomatoes, blanched, peeled and chopped
500g/18oz lean pork, cut into 2.5cm/1in cubes
275g/10oz okra or canned lima beans

KALUA PIGS

This particular pig is stuffed with sizzling hot rocks before being lowered into the pit and surrounded by yams, taro and breadfruit. A thick layer of burlap, canvas and earth seals it in until it is cooked.

1 medium-sized bitter melon, seeded and cut into 1.5cm/½in slices or 200g/7oz canned bitter melon, drained and cut into 1.5cm/½in slices (optional)
shrimp paste or thin soy sauce
plain boiled rice, to serve

- ***Serves 6*** 
- ***395cals/1660kjs per serving***

1 Heat the oil in a large, heavy saucepan or flameproof casserole over medium heat. When hot, sauté the garlic until light brown, stirring constantly. Turn the heat to medium-low, add the onion and ginger and cook until the onion is soft, about 3 minutes. Cut the aubergines and sprinkle lightly with salt. Allow to stand for 30 minutes.

2 Add the tomatoes, pork and 300ml/½pt water. Bring to the boil, lower the heat, and simmer in the covered casserole for 50 minutes.

3 Add the trimmed okra, drained aubergine, bitter melon, if using, shrimp paste or soy sauce to taste. Return the stew to the boil, lower the heat, cover, and continue to simmer, stirring occasionally, until the pork is tender and the vegetables are cooked, about 15-20 minutes. Adjust the seasonings and serve accompanied by plain boiled rice.

Kapalua butterfly

This refreshing cocktail was invented by the bartender of the Kapalua Bay Resort hotel on the island of Maui.

- ***Preparation: chilling the glass, plus 10 minutes***

3tbls dark rum
2tbls unsweetened pineapple juice
2tbls freshly squeezed orange juice
1tbls canned coconut syrup
1tbls fresh lemon juice
about ½tsp sugar
1-2 large fresh pineapple chunks
dash of grenadine
crushed ice

- ***Serves 2*** 
- ***290cals/1220kjs per serving***

1 Combine all the ingredients in a blender, and blend until smooth.

2 Pour the mixture into tall, chilled glasses and serve immediately.

Hawaiian fruit cups

- ***Preparation: 30 minutes, plus chilling***

1 large ripe pineapple
2 small ripe papayas, peeled
3 large oranges
400g/14oz canned lichees, drained and halved
4tbls lemon juice
sugar to taste

- ***Serves 6*** 
- ***140cals/590 kjs per serving***

1 Cut the flesh of the pineapple, papayas, and oranges into small pieces of uniform size and put them into a bowl. Add any juice that is released by the fruit to the bowl.

2 Combine all the fresh fruit with the lichees and toss the mixture in the lemon juice.

3 Taste and add sugar if wished. Chill for at least 1 hour and serve in small bowls or in hollowed out half pineapple or coconut shells.

Variations

For a sophisticated version of this recipe, add 2-3tbls of a fruity liqueur, such as Kirsch or Grand Marnier.

VOLCANIC ERUPTIONS

Traditional Hawaiian cooking developed without fireproof containers, the food (usually a huge pig) being cooked in a large pit lined with wood and volcanic lava rock. This is the famous luau feast and the meat is cooked by using a combination of the heat from the wood and steam.

INDEX

ACKNOWLEDGEMENTS

The publishers extend their thanks to the following agencies, companies and individuals who have kindly provided illustrative material for this book. The alphabetical name of the supplier is followed by the page and position of the picture/s.

Abbreviations: b = bottom; c = centre; l = left; r = right; t = top.

Duncan Brown: 112bl. Bruce Coleman: 56tr. Robert Harding: 34tl. Hutchison Library: 10br, 16tr, 100bl, 106bl, 124bl. Image Bank: 44tr, 118bl, 138tr. Impact Photos: 28tr. Marshall Cavendish Picture Library: Bryce Attwell 30b, 50b; Paul Bussell 10tl, 16br, 17bl, 21c, 24bc, 26b, 33c, 36b, 43c, 52tr, 58bc, 61c, 64bc, 67tc, 82br, 83ct, 80bl, 84br, 91br, 92tr, 97bl, 100tr, 103b, 124tr, 128bl, 133c & back cover, 134br, 135tl, 136tl, 142b; Alan Duns 46tl, 48bl, 117c, 118tr, 119bl, 120cl, 122tl, 122br; Ray Duns 35tl; Laurie Evans 18tr, 20tr, 32t, 44br, 65br, 76tr, 79c, 86br, 90tr, 93tc, 98bl, 107tl, 107br, 108tl, 109bl, 132bl, 139br, 140bc; John Hollingshead 27c, 28bl, 29bl, 31tr; James Jackson 9c, 12br, 13bl, 47tr, 60bc, 63br, 66cr, 68br, 69b, 74br, 77b, 78tr, 113t, 114b, 126b, 129c & front cover, 137c; John Kevern 14b, 22br; Chris Knaggs 20bl, 38tl, 57tr, 71bl, 92bl, 96tr, 123c, 126tr, 131tr; James Murphy 99tc, 102tr, 104bl; Peter Myers 15c, 18bl, 34br & back cover, 40ct, 42cr, 49c, 53bl, 55tc, 58tl, 59tr, 73tc & front cover, 81tr, 94b, 101b, 110b, 119tr, 128tr, 130br, 141tr; Roger Phillips 95tr; Paul Webster 19bl, 37tr, 111c & back cover, 116b; Andrew Whittuck 72tr; Paul Williams 11br, 54b, 56bl, 62t, 76bl, 85tl, 87c, 88br, 98tr, 105tc, 109tr, 120b; Peter Williams 39c, 89tr. All small step-by-step pictures, Marshall Cavendish Picture Library. Spectrum Colour Library: 22tr, 70tr, 80ct, 88ct. Zefa: 40bl, 50bl, 62bl, 74tl, 84tl, 94tr, 130tr, 134tl.

Index compiled by INDEXING SPECIALISTS, Hove.